Richard Suchenwirth
Jun Guo
Irmfried Hartmann
Georg Hincha
Manfred Krause
Zheng Zhang

Optical Recognition of Chinese Characters

Advances in Control Systems and Signal Processing

Editor: Irmfried Hartmann

Volume 8

Volume 1: Erhard Bühler and Dieter Franke
Topics in Identification and Distributed Parameter Systems

Volume 2: Hubert Hahn
Higher Order Root-Locus Technique with Applications in Control System Design

Bernhard Herz
A Contribution about Controllability

Volume 3: Günter Ludyk
Time-Variant Discrete-Time Systems

Volume 4: Dietmar Möller/Dobrivoje Popović/Georg Thiele
Modeling, Simulation and Parameter-Estimation of the Human Cardiovascular System

Volume 5: Günter Ludyk
Stability of Time-Variant Discrete-Time Systems

Volume 6: Irmfried Hartmann/Werner Lange/Rainer Poltmann
Robust and Insensitive Design of Multivariable Feedback Systems
– Multimodel Design –

Volume 7: Ulrich Kramer
Driver Performance Modelling

Volume 8: Richard Suchenwirth/Jun Guo/Irmfried Hartmann
Georg Hincha/Manfred Krause/Zheng Zhang
Optical Recognition of Chinese Characters

Richard Suchenwirth
Jun Guo
Irmfried Hartmann
Georg Hincha
Manfred Krause
Zheng Zhang

Optical Recognition of Chinese Characters

With 73 Figures

Springer Fachmedien Wiesbaden GmbH

CIP-Titelaufnahme der Deutschen Bibliothek

Optical recognition of Chinese characters /
Richard Suchenwirth ... [Ed.: I. Hartmann]. –
(Advances in control systems and signal processing; Vol. 8)
ISBN 978-3-528-06339-9 ISBN 978-3-663-13999-7 (eBook)
DOI 10.1007/978-3-663-13999-7

NE: Suchenwirth, Richard M. A. [Mitverf.];
Hartmann, Irmfried [Hrsg.]; GT

Editor:

Dr.-Ing. I. Hartmann
Prof. für Regelungstechnik und Systemdynamik
Technische Universität Berlin
Einsteinufer 17 - EN11
1000 Berlin 10, West Germany

Ursprünglich erschienen bei Friedr. Vieweg & Sohn Verlagsgesellschaft mbH, Braunschweig 1989

ISSN 0724-9993

ISBN 978-3-528-06339-9

Contents

Acknowledgements

This book was produced in the Joint Research Project IFP 1074 TECHIS (Teilautomatische Erkennung von chinesischer Schrift = part-automatic recognition of Chinese characters) which is being conducted since October, 1986 by the institutes of Measurement and Control Techniques, Communications, and Linguistics of the Technical University of Berlin. This project is explicitly funded by the University. Wilfried Adam, Peter Cassiers, Detlef Schücker, Su Wen-Liang and Sun Bingying have substantially contributed to the project. Wilbur Jobe spent many days polishing the style of this book, a task especially difficult as no one of the authors is a native speaker (or writer) of English. The text portions of the present book were produced by desktop publishing, but without Su Wen-Liang's diligent cut-and-paste work at the real desk, it would have looked much worse. Thanks to you all!

0 Introduction

Optical character recognition (OCR) is an input technology that has been researched on digital computers since the 1960s. It aims to automatically recognize characters: letters, numbers, special characters, whether handwritten or printed. (Of course, the wider variation of handwritten characters poses more problems than uniformly printed characters do). The recognition result is returned as a coded representation of the input characters. These character codes may be used for all kinds of computer processing, from storage in databases to machine translation. In these applications, an OCR device thus performs the function of a keyboard or other data entry devices.

The optical recognition process can be divided generally into the stages of

- optical input of the original text;

- preprocessing (including digitizing and segmentation into separate characters);

- feature extraction;

- classification (candidate selection);

- output of character codes.

Each of these phases will be devoted a chapter in this book.

Typical OCR systems, as they have matured over many years of research and are now being produced and used in increasing numbers, work on limited fonts that may for example include alphanumerics (letters, numbers) and punctuation characters, in some cases even only numbers. Current research concentrates on the more difficult tasks, one of the most demanding being the recognition of Chinese characters.

Chinese writing is the world's oldest writing system still in use today. In spite of its numerous difficulties (a character set that can only be measured in thousands, the majority of which are much more complex than alphanumerics), it is still used today for most writing and reading in China. In Japan and South Korea, Chinese characters still form an important sub–system of the national writing systems. Many years ago, the question was debated whether the obvious impractibility of this gigantic system would force its abolition in favor of the Latin alphabet – one reason being that Chinese characters could be processed only with tremendous difficulties in office equipment like typewriters or computers.

Things have changed since then. Although in Chinese bureaux most correspondence is still written by hand, the software and hardware technologies have progressed so fast that Chinese character processing can be implemented on small, inexpensive computers with sufficient memory capacity and output resolution. The last frontiers for research in this field are acoustic recognition of the spoken Chinese language - and the topic of this book, the optical recognition of Chinese characters.

1 Chinese Characters: Properties and Problems

1.1 History

Most of the oldest known writing systems (the Sumer glyphs, that later evolved into cuneiforms, after c. 3700 BC, Egyptian hieroglyphs c. 2200 BC, Creta, Indus and Chinese scripts after c. 2000 BC) have come out of use thousands of years ago. Only the Chinese *Hanzi* (*Han* being an old name for China, *zi* meaning character) managed to survive up to the present day, and thousands of them are still put to wide-spread use in the most populated country of the world (see fig. 1.1-1 for an example). A transition to the incomparably simpler Latin alphabet was the final aim of Chinese writing reform policy since the 1950s, but in recent years it seems to have been discarded. Chinese characters are here to stay for an extended period of time.

The large number of characters used results from the fact that the *Hanzi* never lost their function of directly representing meanings (in most cases correspon-ding to the modern concept of **morphemes**, the smallest linguistic units that still carry a meaning) as well as sounds, typically a syllable. All other writing systems at some time in their history switched to assigning sound classes, phonemes, or syllables to written signs - meanings could therefore be represented only indirectly. Because in any language the size of the phoneme or syllable inventory is far less than the number of morphemes, the size of the character set in these writing systems is thus greatly reduced.

Owing to the previously dominant role of the Chinese culture, a considerable number of *Hanzi* forms also part of the writing systems of Japan and Korea, although these two countries' national languages have no genetic relationship to the Chinese language, making the *Hanzi* usage even more cumbersome. In multilingual applications, like libraries, the *Hanzi* are sometimes referred to as CJK characters (Chinese, Japanese, Korean) [HK 84:9].

1.11 The "liu shu": Six classes of characters

According to traditional Chinese philology, all *Hanzi* may be classified in one of six classes (*liu shu*, "six [kinds of] writing"). This theory, first mentioned around 300 BC in "The Rites of Chou" [Kra 68:148], may with a grain of salt still be used today as a chronological framework for relating the history of the *Hanzi*.

关于疏散星团照相相对自行的一个正确计算法*

周兴海 经加云
(中国科学院紫金山天文台) (中国科学院上海天文台)

提 要

本文在线性模型的假定下，给出了当参考星自行不知道时，如何计算正确相对自行的一个方法，并且证明了用通常方法(即假定参考星自行满足条件: $\Sigma x\mu_x = \Sigma y\mu_x = 0$, $\Sigma x\mu_y = \Sigma y\mu_y = 0$, $\Sigma\mu_x = \Sigma\mu_y = 0$)求得的自行不是相对自行，两者之间的差值跟待测星在底片上的位置有关.

一、

在疏散星团照相自行测定的实践中，通常采用线性模型:

$$\begin{aligned}\Delta x &= ax + by + c + \mu_x\Delta t,\\ \Delta y &= dx + ey + f + \mu_y\Delta t.\end{aligned} \tag{1}$$

其中 Δx, Δy 为两期底片测量坐标 x, y 的差值, μ_x, μ_y 为 x 和 y 方向的自行分量, Δt 为两期底片历元差, a,b,c,d,e,f 为联系常数.

当参考星自行不知道时，为了从方程(1)得到解，需假定参考星自行满足以下条件[1]:

$$\Sigma x_r\mu_{x_r} = 0,\ \Sigma y_r\mu_{x_r} = 0,\ \Sigma x_r\mu_{y_r} = 0,\ \Sigma y_r\mu_{y_r} = 0; \tag{2}$$

$$\Sigma\mu_{x_r} = 0,\ \Sigma\mu_{y_r} = 0. \tag{3}$$

r 表参考星. 所选参考星或者全为背景星，或者全为团星，或者部分背景星，部分团星.

用这样的办法求得的自行(以下称做计算自行)，一般认为是相对于参考星的自行(相对自行). 其实，这只有在条件(2)满足的情况下，方才是正确的；当条件(2)不满足时，所求计算自行不同于相对自行，两者的差值跟星象在底片上的位置有关.

下面给出在条件(2)不满足时，如何计算正确相对自行的一个方法. 为了推导简单，假定参考星在底片上的位置满足条件:

$$\Sigma x_{r_i} = 0 \quad \Sigma y_{r_i} = 0. \tag{4}$$

这个条件虽然对参考星的选择做了限制，但是在实践中是不难实现的.

方程(1)是在测量坐标(x,y)和标准坐标(ξ,η)之间的转换关系为线性，即

$$\begin{aligned}x &= A\xi + B\eta + C,\\ y &= D\xi + E\eta + F,\end{aligned} \tag{5}$$

Fig. 1.1–1 Portion of a Chinese text

日	月	木	女
rì	*yuè*	*mù*	*nǚ*
sun	moon	tree, wood	woman

Fig. 1.1–2 Some pictograms

Many of the earliest Chinese characters were simplified pictures of the concrete objects they meant (*xiangxing zi*, often translated as **pictograms**, "picture signs"; see fig. 1.1–2). By the 12th century AD, about 600 pictograms existed [DeF 84:84].

Abstract concepts have been written with ideograms, "idea signs", ever since the earliest phases. Chinese scholars distinguish **simple ideograms** (*zhi shi zi*, "characters pointing at facts" – fig 1.1–3) whose components' meaning is determined only in the context of the whole character, and **compound ideograms** (*hui yi zi*, "characters where meanings meet"), where two or more existing characters are arranged together to form a new character, its meaning likewise being a compound of the meaning of its components. In the *Shuo Wen Jie Zi*, the earliest Chinese dictionary, 125 simple and 1,167 compound ideograms are listed [DeF 84:84].

一	二	上	下
yī	*èr*	*shàng*	*xià*
one (number)	two	top, on, above	bottom, below

Fig. 1.1–3 Simple ideograms

Typical examples for the latter kind are the character *ming* "bright" consisting of *ri* "sun" and *yue* "moon", or *hao* "good, love" made up of *nü* "woman" and *zi* "child" (see fig. 1.1–4).

明	東	旦	好
míng	*dōng*	*dàn*	*hǎo*
bright	east	morning	good, love

Fig. 1.1–4 Compound ideograms

Note that the three classes of characters mentioned so far are, to a certain degree, **language-independent**. It does not matter whether the word for "hand" is pronounced *shou* in Chinese, *te* in Japanese or *mano* in Italian: these words could be written with the same (Chinese) character in all three languages. Consider also the similar language independence of Arabic numerals.

In Chinese, problems arose with particles or verbs that could not be written with either pictograms or ideograms. In such cases, **phonetic loans** took place: characters with equal or similar pronunciation that had already been assigned a different meaning were used purely as "phonograms" (*jia jie zi*, "falsely borrowed characters"). These characters serve only in a phonetic function and demand that the reader recall the conventions that govern the relation between sound, graph and meaning in the given (Chinese) language.

The more this elegant principle (which after all also was the foundation for modern alphabetic writing systems in the Western world) was employed, the more ambiguities could arise about the meaning of a given character. To remedy this situation, a further category, the **phono-ideograms** (*xingsheng zi*) evolved.

盟	棟	但	刖
méng	*dòng*	*dàn*	*yuè*
union, league	roof beam	but	foot-cutting

Fig. 1.1–5 Some phono-ideograms

These characters consist of two components, one (the "radical", *bushou*) giving a more or less vague hint at the semantic category, the other (the "phonetic", *shengpang*) serving as indicator to the pronunciation in the same way the phonetic loan did (see fig. 1.1–5). The position of the radical is fixed only by convention: it may be placed to the left or to the right, above or below the phonetic; in some cases, the radical encloses the phonetic, while in other characters it is placed inside the phonetic.

The number of radicals differs in various approaches at systematization. In the oldest dictionary *Shuo Wen Jie Zi*, 540 radicals were used. This number was reduced in the dictionary *Zihui* (1615) to 214. Because they were made famous by the *Kangxi Zidian*, these are mostly called *Kangxi* radicals by contemporary sinologists. This system is still the standard for dictionary arrangement in Taiwan, Hong Kong and Japan. In the PRC, various shorter and longer lists have been experimented with, but no standard exists anymore.

Although phono–ideograms created in that way tend to be more complex than characters of all other categories, the advantage of having indicators for both meaning and pronunciation made the phono–ideographic principle most popular for coining new characters: While in writings from the Shang dynasty (16.–11. century BC) phono–ideograms accounted for just about 34 % of the character set used, this proportion grew to 97 % by the year 1716 when the gigantic dictionary *Kangxi Zidian*, containing over 47,000 different characters, was published [DeF 84:84]!

Careful readers may have noticed that so far only five categories of characters have been discussed. The missing class, *zhuanzhu zi*, was traditionally defined as "one character circularly defining the other". Examples are very rare. Most writers on Chinese characters prefer to neglect this class, and we will do the same here.

1.12 Styles: From Oracle Bones to Movable Types

It comes as no surprise that over a period of more than 3,000 years, the technologies used to write Chinese characters as well as the writing styles underwent changes. The earliest archeological findings that can be counted as Chinese characters are scratched on animal bones and turtle shells (hence the name ***jiaguwen***, "shell/bone script") that were used in forecasting the future. The bones or shells were inscribed with the questions and thrown into a fire, which made the material develop cracks. The diviner later interpreted these cracks and issued his prognoses that were also written on the turtle shells.

Similar characters (which by the way were almost as developed as today's and existed in many variants) can be found on bronze vessels and tripods (***jinwen***, "gold (i.e. metal) script"). They were cast with the vessel and bore notes on sacrifices or laws to be publicized. These two forms are usually referred to collectively as ***dazhuan***, "big seal script", to distinguish them from the first known standard form, the ***xiaozhuan*** ("small" or "lesser seal").

This modified set of characters was defined by Chancellor Li Si as part of the reforms he introduced following China's first unification in historical times, the founding of the Qin (Ch'in) dynasty by Emperor Qin Shi Huangdi in 221 BC. Li Si weeded out variant writings and in some cases severely changed the character forms, leading Western scholars to the conclusion that the small seal "is in many cases an entirely new script" [Kar 62:49].

A very comprehensive set of about 10,000 *xiaozhuan* characters, together with explanations on etymology and meaning, is found in the *Shuo Wen Jie Zi*, the venerable dictionary compiled around the year AD 100 by Xu Shen and reprinted until today.

It was also in the Han Dynasty (206 BC–AD 220) that the writing tools gradually changed: the bamboo "pen" that produced lines of uniform width was replaced by the animal hair brush. By exerting more or less pressure on the brush, stroke thickness could be freely varied, but not all curves could be drawn graciously. This led to a change in writing style: in the ***li shu*** ("chancellery style") known since the end of the Qin dynasty (207 BC), round turns were frequently replaced by broken corners.

After a number of relatively slight modifications, *li shu* evolved into the ***kai shu*** ("regular writing") style, the creation of which is traditionally attributed to Wang Cizhong (c. AD 80) [Kar 62:49]. Until today, *kai shu* represents the standard style of *Hanzi* for both hand-written and printed characters ever since Bi Sheng invented printing with movable (clay) types in the year 1040, 400 years before a certain Johann Gensfleisch zur Laden, a.k.a. Gutenberg.

Frequent use of the ink brush led the scribes to take shortcuts. Ever since the Qin/Han period, they joined sequences of several distinct strokes into one, sometimes still retaining the *gestalt* of the character in question (this style is known as ***xing shu,*** "running hand"), sometimes deviant to such a degree that the new written form bore little resemblance with the *kai shu* original (this style is called ***cao shu,*** "grass writing"). Any individual hand might, however, float freely between the extremes of clean, print-like writing and almost illegible scribble: "A scholar almost considered himself disgraced if he wrote a readable hand like a common scribe" [Kar 62:51].

For decorative purposes, all these writing styles from the Lesser Seal downward are still used today in calligraphy (see fig. 1.1–6), and examples of each style are to be seen anywhere in China on shop signs or magazine mastheads.

Fig. 1.1–6 Various styles: kai shu, li shu, xing shu, cao shu, seal script

1.13 Applications: The "Latin of East Asia"

From the 4th century AD to the latter half of the 20th century, the Chinese culture exerted a tremendous influence on most of the neighboring countries in East Asia: the traditional education systems of Korea, Japan and Vietnam were firmly based on the Chinese system with its three philosophical or religious streams of Confucianism, Buddhism and Taoism.

The Classics or Holy Books of these doctrines were written in (or, in the case of Buddhism, translated into) the Classical Chinese language (*wenyan*), so for every educated person the mastery of the Chinese language and writing system was an absolute prerequisite to any learning.

This situation, to a certain degree, parallels the development in mediaeval Europe, where Latin was the "cultural" language used by scholars, monks and every educated person for writing. A second and even stronger parallel lies in the fact that a host of technical and other terms from the "high level language" (Latin/Chinese) were borrowed and firmly incorporated into the "vernacular" languages – sometimes to such a degree, like in English or Japanese, that the imported words (especially nouns) amount to a third to half of the total set in the receiving language.

The major crisis of the traditional Chinese culture broke out in the 19th century, when European powers came into closer, and often unfriendly, contact with the Middle Kingdom (as the Chinese word for China is literally translated), a country that up to then had only known culturally inferior, "barbarian" nations other than itself. Western culture was new in at least having far superior weapons technology as well as the required scientific and technical foundations, and an urge to "learn from the West" emerged in most East Asian countries.

The strongest impulse was notable in Japan, where since the Meiji reform of 1868 huge amounts of Western books were read – and translated, leading to the coining of a large number of new terms in the Japanese language. The material to build these new terms were however still the *Hanzi*, linked together in new combinations that looked like words from Classical Chinese. Here we see another parallel between Latin and Chinese: when creating new technical terms, it is not surprising for an Englishman or German to combine Latin (or Greek) morphemes into a new word – that later may very well be borrowed into Italian and be used in Rome, where all Latin originated after all. A similar situation is found also in the Modern Chinese language: a considerable number of terms has been borrowed from Japan, where in turn they had been composed of Classical Chinese morphemes.

In China's three neighboring countries (or rather regions), there existed no writing system before the Chinese characters. It was only after exposition to the *Hanzi* system that local scholars started to develop scripts for recording the vernacular languages, that were all used together with *Hanzi*:

In **Korea**, the *Idu* characters from the 7th century, and in use until the 19th century, were actual *Hanzi* stripped of their meanings, i.e. all used as phonetic loans. As the Korean language contains syllables that do not exist in the Chinese language, finding adequate *Hanzi* was often difficult, if not impossible. This led in 1443 to the design of the *hangul* or *onmun* script, an alphabetic writing system with signs for vowels, diphtongs and consonants that were arranged horizontally and vertically (resembling very much the structure of *Hanzi*) to form square compounds, one for each syllable. See [Bec 85] for more details.

Japan received the first lessons of the Chinese culture in the 4th century AD via Korea: Buddhist monks crossed the sea to promote their religion and their writing culture among a people that, like the Koreans before, had no writing system at all. So they eagerly accepted *Hanzi* (or, as they were called in Japan, *Kanji*) as the standard of writing, and *wenyan* as the standard language of education. Beginning in the 9th century, the *wenyan* written in Japan began to move closer to the Japanese language of that time: grammatical features, most notably the final position of the verb, as well as Japanese words were integrated in the evolving Classical Japanese (*bungo*). To write down vernacular words that were not defined in Chinese, the phonetic loan principle was again applied; in some cases, specific *Kanji* (*kokuji*, "national characters") were newly coined that look like Chinese characters but cannot be found in any Chinese dictionary.

In ever more frequent usage, the contours of the *Kanji* used phonetically were written more and more cursively (in parallel with the "grass writing" style that was already popular in China, see 1.12) and thus simplified. It was only in the Edo period (1603–1868) that Japanese scholars standardized the set of *kana* ("borrowed names") and arranged them to form the 50-syllable table of the modern Japanese phonetic script, *hiragana*. The other set of *katakana* syllable signs is similarly organized, the character forms were derived from the original *Kanji* by retaining only the first characteristic strokes.

Chinese characters are, however, still used in Japanese and some South Korean texts when writing loan words form Chinese, leading to the typical mix of fonts. See [Bec 85:28] for some examples. There are tendencies to simplify the writing system by abolishing Chinese characters (as was done in North Korea in 1948), but for instance in the development of word-processing computers in both Japan and South Korea, great care is always taken to include a couple of thousand Chinese characters in the system's character set.

The development in **Vietnam** was different in that a writing system for the broad public and the vernacular language was developed by European missionaries in the 17th century, and therefore was based on the Latin alphabet. But from the 8th century AD, Vietnamese words embedded in *wenyan* texts had been written with specially coined characters (*chu nom*) that looked like *Hanzi* but weren't. Typically, a Chinese character with the meaning of the Vietnamese word to be written was combined with a (sometimes second) phonetic that resembled the pronunciation of the Vietnamese word: in contrast to the practice in Korea and Japan, not the phonetic loan but the phono-ideographic principle was employed. The *chu nom* were, however, used only by a relatively small class of scholars and maybe justifiably died out in the mid-20th century. Since Vietnamese is one of the languages typologically most closely related to Chinese, the Vietnamese transition to alphabet writing is often cited as an argument that latinization is also feasible in China.

1.14 Character Simplification

The effort needed to write or learn a character is related to its complexity, which for *Hanzi* is measured typically by the number of strokes. In hand-written characters, especially in the "grass style", there has been a tendency for centuries to simplify characters by reducing their complexity. The forms of printed characters, however, were for a long time left unchanged and uniform throughout the areas where *Hanzi* were used.

traditional	東	佛	應	舊
Japan	東	仏	応	旧
PRC	东	佛	应	旧

Fig. 1.1-7 Examples of character simplification

This situation began to change after World War II. In 1949, the Japanese Government decreed the official simplification of 320 printed characters. From 1956 to 1964, the People's Republic of China (PRC) followed suit by introducing more than 2000 simplified characters, which however were only partially compatible with the Japanese variants (of the 320 Japanese simplifications, 183, or 57 %, were simplified differently in the PRC!). The number of PRC simplifications cannot be stated exactly since the official lists also contain a number of radicals or components that are to be simplified in all characters in which they form a part. [DeF 84:260] states that 2,238 out of the approx. 7,000 characters in general use had been simplified by the year 1964. The average stroke count in running text was reduced by 16.1 %, from 9.15 to 7.67 strokes. For examples, see fig. 1.1-7.

A "Second List of Simplified *Hanzi*" was published in major PRC newspapers in December 1977, containing around 250 resp. 300 simplifications in two parts. The characters from part 1 were subsequently employed in printing newspapers and some books, but after half a year they were withdrawn. The characters in part 2 were never used in print.

Although the list of 1977 has been cancelled and printed characters again conform, or are at least supposed to conform, to the standard of 1964, some of the 1977 simplifications can still be observed in handwriting and lettering. For the near future, no further simplification of characters is planned in the PRC. The Committee for Writing Reform, *wenzi gaige weiyuanhui*, even changed its name to "Committee for Language and Script Work", *yuyan wenzi gongzuo weiyuanhui*, thus doing away with the term "reform".

In **Singapore**, a number of printed characters were simplified in 1973, some of them neither compatible with Japanese nor PRC simplification. From about 1979, these characters were changed again to correspond with the PRC simplifications.

No simplification of printed *Hanzi* took place in **Taiwan**, **Hong Kong** and **Macau** as well as in most publications of overseas Chinese (in Europe and America). In these areas, the traditional character forms are still the standard.

1.2 Modern Printed Characters

Printing with movable types was invented in China around the year 1040, but the clay types of then were never widely used. Most pre-modern books were printed from wooden blocks, one block hand-carved for each page. Printing with lead types was re-introduced from the West in the second half of the 19th century and is still in use today, although major printshops, like at the *Renmin Ribao* (People's Daily), have already switched to photo or laser typesetting.

1.21 Font Styles

Type foundries or rather their modern successors in Western countries offer a bewildering variety of thousand or more different fonts. It may be because of the enormous size of the character set that for Chinese characters no similar richness of fonts exists. In printed books, magazines and newspapers, four major categories of font styles are used in China. To be sure, there are slight to considerable variations of character instances inside one font style that for Western typographical eyes would justify a distinction between different fonts. In China and Japan, these differences are neglected: Only the broader categories of font styles, but no individual fonts are named.

The most frequent font style used for body text as well as some headlines is called ***songti*** ("Song font", after the Song dynasty, AD 960–1279) in China or ***mincho*** ("Ming dynasty", 1368–1644) in Japan. It may boldly be compared to Antiqua fonts in the Latin alphabet: horizontal strokes carry a triangular ornament (a kind of serif) in the top right corner and are thinner than vertical strokes.

The ***fangsongti*** ("imitated Song font") has livelier strokes that are not as geometrically constructed as in *songti*. Horizontals are slightly slanting upwards. This font is used for subheads in print, not so often for body text. It is also the typical font of Chinese mechanical typewriters.

The ***heiti*** ("black font"), or in Japan ***gosikku*** ("Gothic"!), may be likened to demi-bold sans serif fonts in alphabets. Horizontals and verticals are of equal thickness without serifs. This font is mostly employed for headlines or specially marked portions of text.

The last category, ***kaiti*** ("standard font"), bears resemblance to italic or script fonts in the Latin alphabet. It is modelled after *kai shu* handwriting and has even more lively shapes than *fangsongti*. *Kaiti* is typically used in prefaces, subheads and footnotes, sometimes also to mark names of persons inside *songti* body text.

(42 P) 全面开创社会主

(28 P) 全面开创社会主义现代化

(21 P) 全面开创社会主义现代化建设的

(21 P) 全面开创社会主义现代化建设的

(16 P) 全面开创社会主义现代化建设的新局面

(16 P) 全面开创社会主义现代化建设的新局面

(16 P) 全面开创社会主义现代化建设的新局面

(16 P) 全面开创社会主义现代化建设的新局面

(14 P) 全面开创社会主义现代化建设的新局面

(14 P) 全面开创社会主义现代化建设的新局面

(14 P) 全面开创社会主义现代化建设的新局面

(10.5 P) 全面开创社会主义现代化建设的新局面

(10.5 P) 全面开创社会主义现代化建设的新局面

Fig. 1.2–1 Standard fonts in various sizes

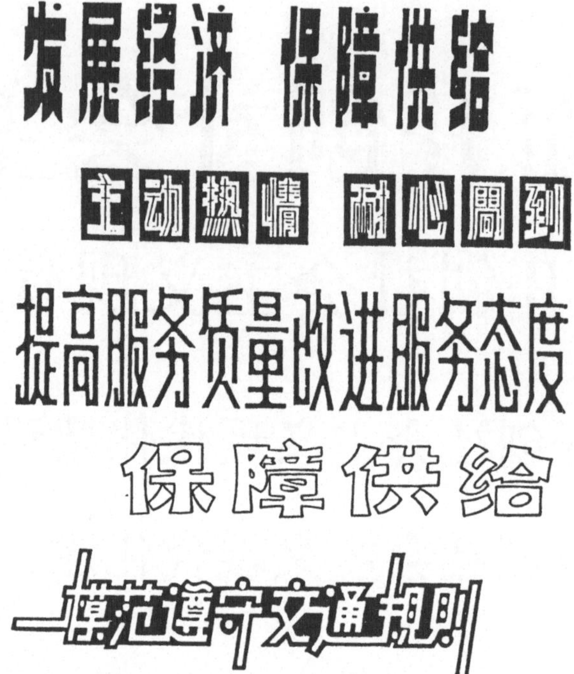

Fig. 1.2-2 Examples of fancy fonts

For examples of these fonts in different sizes, see fig. 1.2-1. Besides the major categories of sober fonts for body text there exists, not unlike the situation in Latin alphabets, a huge number of decorative "headline" fonts that draw freely from all styles in calligraphy as well as from Western design and are used in advertising or magazine headlines (see fig. 1.2-2). In China, these are however mostly hand-lettered.

1.22 Font Sizes

According to a sample from a Chinese printshop (see fig. 1.2-1), the following type sizes (height of the character blocks) are in use:

Chinese Name	Pica	mm
Chu hao	42p	14.8
1-hao	28p	9.8
2-hao	21p	7.4
3-hao	16p	5.6
4-hao	14p	4.9
5-hao	10.5p	3.7
6-hao	8p	2.8

A Pica point is a typographical size measure as used in the United States and Britain. It measures 0.013837 inches, or about 1/72 of an inch. It has to be distinguished from the continental Didot point which is slightly larger [Rub 88:14].

In Chinese printed matter, the *5–hao* and *6–hao* sizes are used for most body text, while the larger sizes are employed for headlines.

1.23 Punctuation and Layout

Traditional Chinese texts consisted of characters only. No punctuation marks were employed in printed texts. Scholars of old used to tick off the boundaries between sentences themselves with the ink brush. Two of these signs, the "inverted comma" *(dunhao)* marking phrase boundaries, and the small circle *(juhao)*, that is used like the Western full stop, have been carried over to modern printed texts. Most other punctuation marks in use today (see fig. 1.2–3) were taken over from Western writing systems.

The traditional **writing direction** was in vertical top–down columns, the columns being written from right to left. Books were also bound at the right–hand edge, so the first page of a Chinese book would be the last page according to Western habits. If occasional horizontal lines were necessary, as for door signs or newspaper headlines, they were set from right to left (although inside characters, the prescribed writing direction is from left to right!). This writing convention is still typical for Hong Kong, Macao and Taiwan.

	Chinese name	Function
○	*juhao*	full stop
、	*dunhao*	links parallel phrases of identical grammatical type (nouns, adjectives)
《...》	*shuminghao*	brackets for names of books, articles etc.
·	*jiangehao*	used between numbers in dates or foreign names
,.:; ? !"		as in Western usage

Fig. 1.2–3 Chinese punctuation marks

In the PRC, this traditional style is also sometimes used. Most printed matter is arranged, nevertheless, in the Western manner: horizontal lines from left to right, the lines running top–down. This style was introduced in the 1950s as one of the first measures of writing reform. It facilitates the embedding of text portions in foreign languages, or mathematical formulas, and was also meant as a preparatory habituation for the planned shift to the Latin alphabet, which, however, seems to become less and less likely these days.

The characters are set tightly without adding extra space at word boundaries. Words may be split at line or page breaks, but no hyphens are used. In the older typesetting tradition still alive in Taiwan and Hong Kong, line or page breaks may even occur before punctuation marks, resulting in lines or even pages that begin with a comma or full stop. In modern PRC texts, the blank space following punctuation marks is justified to prevent these situations.

1.3 Character Structure

Chinese characters vary in complexity from just one horizontal stroke to 36 or more strokes, all arranged in the same square area. From another viewpoint, every *Hanzi* consists of one or more components. Risking the possibility of contradiction, one might say that elementary strokes are the "atoms" forming Chinese characters in that they cannot be subdivided any further, while components may then be likened to molecules that occur separately or again form larger compounds. The structure of the characters will consequently be discussed here from both directions.

笔画 Strokes	名称 Names
丶	点 diǎn
一	横 héng
丨	竖 shù
丿	撇 piě
㇏	捺 nà
㇀	提 tí

笔画 Strokes	名称 Names
乛	横钩 hénggōu
亅	竖钩 shùgōu
㇂	斜钩 xiégōu
㇕	横折 héngzhé
㇄	竖折 shùzhé

Fig. 1.3–1 One list of stroke types

1.31 Strokes

In handwriting, a stroke *(bihua)* is defined as the trace left on paper by one movement of the brush between lowering and raising. This means that sequences of horizontals and verticals may also be counted as only one stroke if the end point of one segment coincides with the starting point of the next segment. Counting strokes in this canonical way is important for dictionary lookup. Recognition systems may need to analyze characters down to stroke elements that contain no abrupt changes in direction [Lu 87].

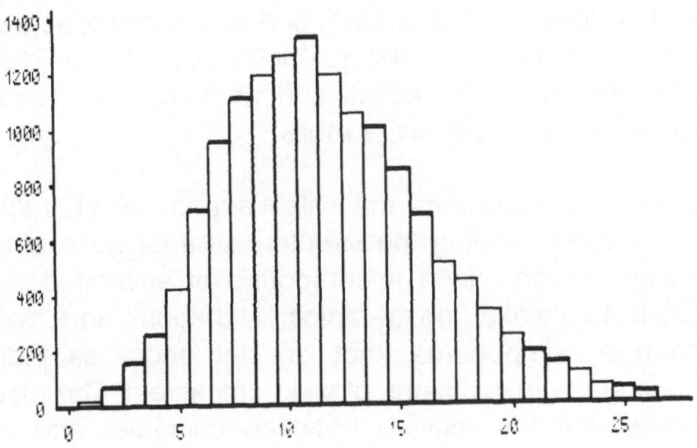

Fig. 1.3–2 Distribution of stroke numbers

DeFrancis [DeF 84:74f] classifies strokes in three groups: dots, lines and hooks, with a further separation according to compass directions. Because of tradition and the pecularity of the writing brush, the directions North, North–West and West are never used, leaving five basic directions possible.

Chinese textbooks give lists of from 5 to 11 basic strokes, while in discussions of calligraphy up to 36 types of strokes are distinguished. One typical stroke inventory is shown in fig. 1.3–1.

The frequency distribution of stroke types was determined by Nagao [Nag 80]:

Horizontals *(heng)*	30 %
Verticals *(shu)*	26 %
Dots *(dian)*	16.5 %
South–West slanting *(pie)*	18 %
South–East slanting *(na)*	9.5 %

In a different count undertaken by Yajima et al. [Yaj 81], 40 % were given for horizontal and 20 % for verticals.

The **number of strokes** may be used as an easy measure of character complexity. Fig. 1.3–2 shows the distribution of stroke numbers among the 14,424 simplified characters in the *Cihai* dictionary (Hong Kong/Beijing 1965). Note that most characters have eleven to twelve strokes.

1.32 Components

Every *Hanzi* consists of one or more components. From a philological point of view, the number of components is one for pictograms and simple ideograms (the character is identical with the component), while phono–ideograms typically consist of two components (radical and phonetic), and composite ideograms of at least two components. Since, however, the phonetic of a phono–ideogram may itself be a character composed of radical and phonetic, the number of components at the lowest level may reach six or more.

The overwhelming majority of components may also appear as characters in their own right. This makes understanding the structure of a Chinese character much easier for the trained person, since a compound of several dozens of strokes can be structured in terms like "nose–radical", "dragon" and "four–dot fire". Many of the non–free components, that do not occur as individual characters, are actually only graphic variations of valid characters. The distortion may go to such a degree that the relation between character and related component shape just has to be memorized without any visual clues.

There is no common opinion on the number of possible components. In various approaches to character analysis or synthesis, different teams have arrived at lists of between 245 and c. 4,000 components. This is in part due to the fact that a number of components may be geometrically subdivided into sub–components without philological reason.

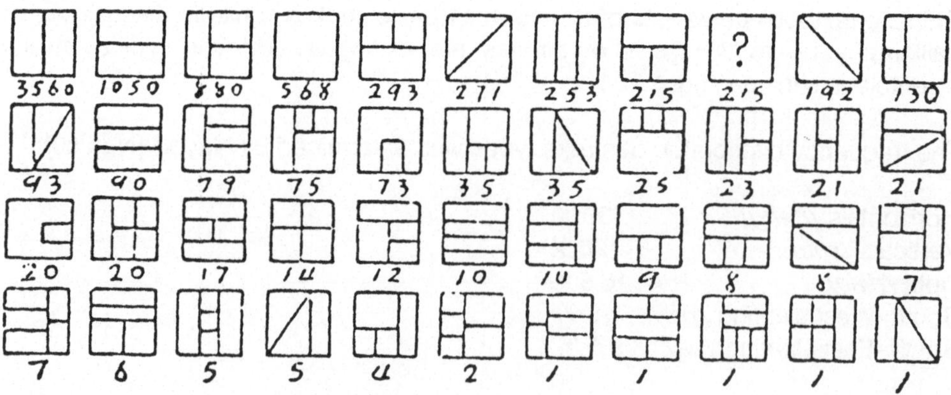

Fig. 1.3–3 Composition structures

Various frequency counts have been undertaken on components, but their results depend strongly on the underlying component set. If a (in one count) frequent component is further divided in another count, the frequencies of these sub-components will also be raised considerably, compared to the first count.

1.33 Composition Structure

There are three basic principles for arranging two components to form a character: left-right (64 %), top-bottom (19 %) and inside-outside (13 %; the percentages were evaluated from 7,254 simplified characters in the dictionary *Xinhua Zidian*). Since a component may again consist of sub-components, more complex composition types can be constructed recursively.

Wei Juxian [Wei 61] determined 43 different composition patterns in an analysis of 8,366 non-simplified characters from the dictionary *Guoyin Xuesheng Zihui* (see fig. 1.3-3). From his results, the following distribution of component numbers can be calculated:

1	568	6.78 %
2	5,166	61.75 %
3	2,031	24.28 %
4	340	4.06 %
5	45	0.54 %
6	1	0.0012 %

This shows that the vast majority of Chinese characters consists of two or three low-level components. See fig. 1.3-4 for some complex characters.

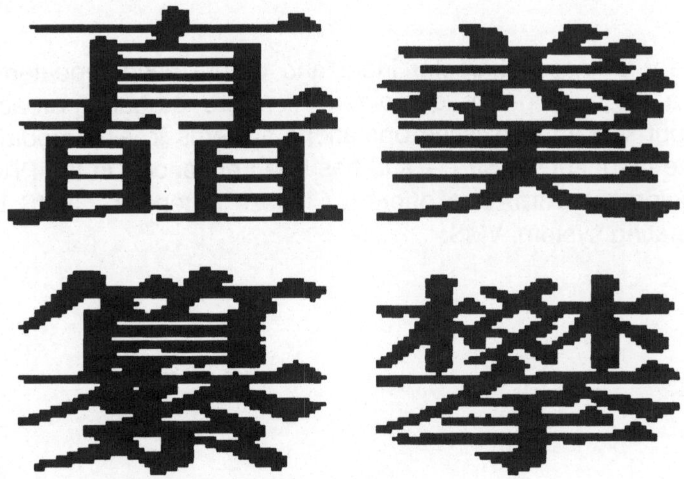

Fig. 1.3-4 Rather complex characters

1.4 Chinese Characters in the Computer

All computing may be divided into the three fundamental phases of input, processing and output. The input of Chinese characters is concerned with transforming a *Hanzi* into the corresponding code, since internally computers can only manipulate sequences of bits. Various input schemes have been developed for Chinese characters. Some require the operator to do most of the coding mentally and key in the result on regular or special keyboards. Other input methods employ character recognition: either optical character recognition (to which the rest of this book will be devoted) or on-line recognition of handwritten characters.

The internal processing of Chinese characters does not differ much from text processing in Western languages, the main problem here being sorting - while in Western languages the arrangement of the alphabet is unequivocally standardized and universally applied, no similarly stringent sorting orders exist for Chinese characters. Just about every new dictionary published in the PRC since the 1950s employs its specific sorting and indexing system, forcing the user to adapt himself to the system currently used.

Output of Chinese characters is mostly done on raster-oriented graphic devices (displays, printers). Vector graphics output to vector screens and plotters was implemented several times in the 1960s and 1970s, but seems to have grown out of fashion alongside with the vector-oriented devices. Segment displays have been experimented with, but met even less popularity, leaving the bulk of applications to raster devices. The only physical requirement is for a reasonably high resolution of the device, which is generally delivered by modern equipment, though.

Application programs for Chinese character input and output have repeatedly been developed for a huge range of computers. A more global approach modifies the input/output system of existing operationg systems to accomodate Chinese characters: the ubiquitous MS/PC-DOS has been enhanced in the PRC to CCDOS. Digital Equipment Corp. also offers a Chinese Extension CVMS to the VAX supermini operating system, VMS.

1.41 Processing Problems

Compared to all other writing systems, Chinese characters pose a number of special problems for technical, especially computer processing: a large set of sometimes extremely complex characters has to be managed, of which many characters may never be used in the system's lifetime, while on the other hand the possibility of out-of-set characters must always be taken into account.

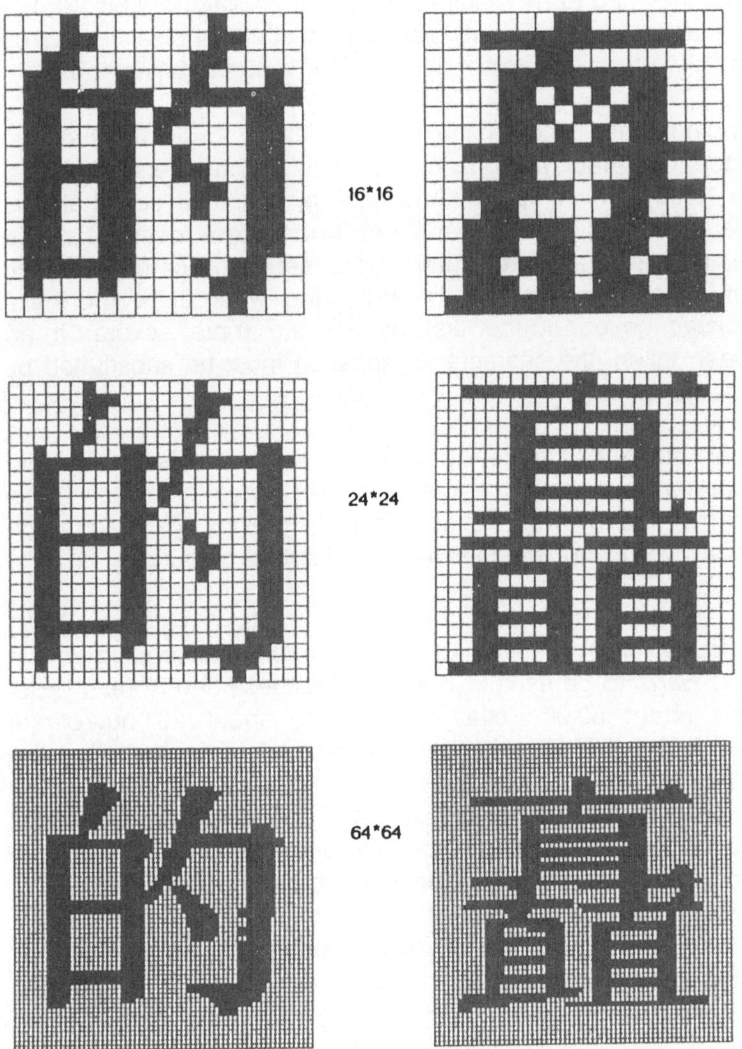

Fig. 1.4–1 Characters in different matrix sizes

As shown before, Chinese characters may consist of 36 strokes or more, or up to six components that themselves can be rather complex. This requires a higher resolution from matrix–oriented output devices (displays or printers): while for alphanumeric characters a matrix size of 5*7 dots suffices for human recognition, the **minimum matrix** required for Chinese characters output is 14*15 dots. Even in this comparatively high resolution, it is not possible to represent all the strokes of some complex characters. In these cases, the matrix image has to be simplified to allow at least the general appearance of the character to be recognized. In a font of 6,763 Chinese characters, 138 characters were so complex that they had to be modified for 16*16 display [Wid 88:37]. For acceptable display or print quality, a matrix size of at least 24*24 pixels is reqired [Mat 85:43]. See fig. 1.4–1 for examples of characters in diffeent matrix sizes.

An even greater problem is the fact that it is hardly possible to delimit "the" set of Chinese characters. Different dictionaries list different numbers of *Hanzi*, starting from about 7,000 to over 50,000. Any working character set is always exposed to two dangers: first, that **out–of–set characters** demand to be processed; second, that a considerable part of the set is never used, which results in waste of storage space and the preparing work done for these characters. The consequences of the first danger are more severe: if no precautions have been taken, the character in question must be substituted by another one or altogether skipped.

In order to adequately solve the problem of out–of–set characters, the system must allow the operator to extend the character set when necessary. This process needs to include the following actions: linking a previously unused code to the new character as well as updating the input and output resources.

A case study: the American Research Libraries Information Network (RLIN) implemented a computerized system for East Asian books cataloguing with a working set of 14,063 *Hanzi* to be used in Chinese, Japanese and Korean texts. After one year and about 50,000 titles catalogued, about 20 out–of–set characters had been found [HK 84:11]. The RLIN procedure for updating the character set stipulates that out–of–set characters are to be coded locally at the terminal where they were found. The coded data are transmitted to the central computer for integration into the existing resources, and the updated resources are in turn loaded down to every terminal in the morning boot procedure.

In another application at the Agricultural Science Information Center (ASIC) in Taipei, the abstract database required a character set of about 30,000 [HK 84:11].

To make things worse, the **occurence frequencies** of Chinese characters are extremely unevenly distributed. In a recent frequency count over N = 1,808,114 character occurrences done in Beijing 1985 [Xia 86], 4,574 different characters

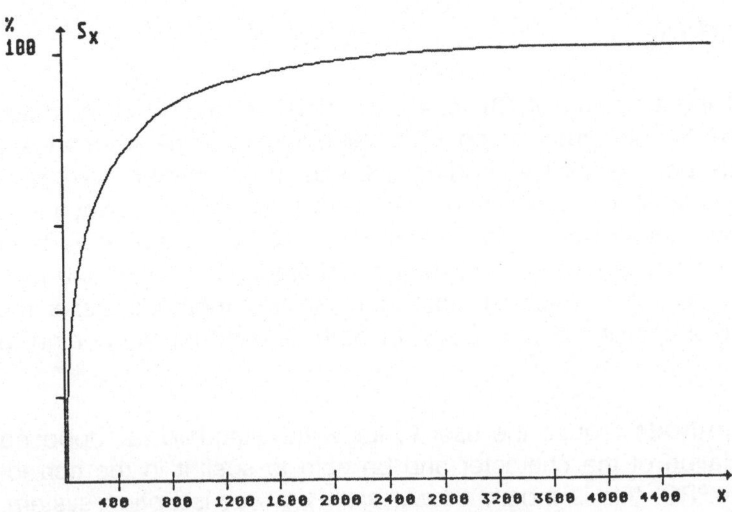

Fig. 1.4–2 Cumulated character frequencies

were found (see fig. 1.4–2). The most frequent character (the structural particle *de*) occurred 75,306 times and thus constituted 4.16 % of the texts analyzed. On the other hand, 425 characters occurred only one time each (.000 06 %).

When larger corpora are examined, the disproportion increases, since there is always a set of "one–occurence" characters with frequencies of 1/N each. In a count done in 1975/76, a total of 21,636,809 occurences was analyzed, yielding 596 "one–occurence" characters with frequencies of .000 004 62 % each.

Considering a text system on which in eight hours every day 10,000 *Hanzi* are processed, the estimated average distance between two occurences of the most frequent character would be 1.3 minutes, while there would be 14.3 years between two occurrences of any of the 596 "one–occurrence" characters! The computer system's lifespan is usually shorter than this estimated period, so characters in this low frequency set very probably will never have to be processed. On the other hand, their occurence can not be excluded. There is no fixed outer boundary to the *Hanzi* set in use, so extremely rare characters may unexpectedly show up, for instance in names of places or persons.

1.42 Input Coding

The oldest form of input coding of Chinese characters, abstract coding, dates back decades before the computer: since 1871, telegrams in China are manually coded in four digits per *Hanzi*. This coding scheme, the Standard Telegraph Code (STC), puts the least requirements on hardware (only a decimal keypad is required) but is most demanding on the operator who has to learn hundreds or thousands of numeric codes for the most frequent characters by heart and look up the codes of the remaining ones. In most modern input methods, either the pronunciation or the shape of the characters (or both) is used for the coding to facilitate learning.

Phonetic input methods require the user to know the standard (as opposed to dialect) pronunciation of the character and be able to spell it in the correct orthography (in the PRC, this is mostly the *Hanyu Pinyin* transcription system, sometimes in abbreviated form [She 85]; outside the PRC, the older *Zhuyin Fuhao* system is sometimes used). These requirements sometimes lead to a generation gap: young educated operators who have learned *Pinyin* in school often prefer phonetic input, while older persons are often frustrated by unsuccessful phonetizations and hence tend to prefer graphic coding.

Another drawback of phonetic input is that the number of syllables in Modern Chinese is very limited (about 410 if tones are not indicated, otherwise still less than 1,400). Frequent syllables like *ji*, *yi* or *yu* correspond to more than 100 different characters! This dilemma is resolved by menu selection (which still can be very time-consuming) or extension of the code to also include hints at the shape of the character. In the most promising systems, input units are not separate characters viz. syllables but words: Chinese words typically consist of two syllables and are much less subject to homophony [Bec 85]. In Japanese input, this is paralleled by the so-called Kana-to-Kanji conversion [Mak 85].

Graphic input methods require the operator to dissect the character into components or strokes, usually in the conventional writing order. In component coding, a set of 26 to more than 1,000 classes of components is defined. Each class is mapped onto a key on the regular (or, with large component sets, especially constructed huge) keyboard. Code collisions do occur, but are much rarer than with phonetic coding. The major ergonomic problem with component coding is that dissecting a character in pre-defined elements is often very tedious and equivocal. Some graphic input methods popular in Taiwan are the Cangjie or Dragon code and the Three Corner Coding Method [Hua 85].

Especially in Japan, **large keyboards** that contain a key for each character are also very popular. These are mostly realized by a touch-sensitive tablet, each key being represented by a small rectangular area that can be touched with a pen. Input rate is about 20 characters/minute [Mat 85]. A Chinese counterpart is depicted in [Woo 85:82].

1.43 On-line Character Rocognition

For persons trained in writing Chinese characters, the easiest computer input method is done by writing the characters with a special pen on a digitizing tablet. When the pen touches the tablet and the in-built switch is turned on, the computer receives data on the pen position in real time. From these data, the recognition program can determine the starting and ending point (as well as intermediate points, if necessary) and thus the direction of each stroke. If the standard stroke order is respected, the sequence of stroke descriptions can easily be matched with corresponding reference descriptions.

The initial attempts by Groner/Heafner/Robinson [Gro 67] aimed at pre-classification: only the direction of the first stroke and the total stroke number were extracted. Fine classification was done manually by menu selection.

At the Japanese Telephone and Telegraph Company NTT, an online input system for *Kanji* was developed in which each stroke is described by a number of feature points (for instance, start, center and end points). After a character is put in, these feature values are compared with the values of all candidate characters of the same stroke number. This method is independent of stroke order (which greatly facilitates its practical use) and achieved recognition results of >99 % over a character set of about 2,300 *Kanji* and *kana* [Mat 85].

On-line input of Chinese characters is well suited for infrequent users because it does not demand learning codes. The **input speed** is limited, however, and cannot exceed the speed of careful handwriting which has been measured to amount to about 800 characters/hour, or 13.3 characters/minute [Zha 79:163].

1.44 Internal and Transmission Coding

Apart from the Standard Telegraph Code mentioned before, Chinese characters are usually represented with a two- or three-byte code for storage and internal processing. When working within a particular computer system, the choice of coding is largely up to the implementer, standardized codes are necessary if data is to be transferred between different computer installations. Such standards are frequently used or modified for internal coding as well.

The first of the modern standard codes was the "Code of the Japanese Graphic Character Set for Information Interchange" (**JIS C-6226**) introduced in 1978 by the Japanese Institute of Standards. This code plan was designed to be compatible with the widely employed ASCII code for alphanumeric characters. The possible values of the byte are limited to those of the 94 printable ASCII characters (hexadecimal 21 to 7E), thereby eliminating control characters that might disturb communications.

The code space can be visualized as a matrix of 94 rows (first byte) with 94 columns (second byte). Next to a number of non-Chinese characters (alphanumerics, Greek and Cyrillic alphabets, *hiragana* and *katakana*), the C-6226 contains codes for 6,349 *Kanji*. This character set has been divided in Level 1: 2,965 frequent characters, arranged "kana-alphabetically" after their major reading, and Level 2: 3,384 rare characters, arranged according to radical and stroke number.

The corresponding national standard of the PRC, the **GB 2312-80** "Code of Chinese Graphic Character Set for Information Interchange. Primary Set", was put in force in May 1981. It is closely modeled after the C-6226, the differences lying mainly in the selection and arrangement of characters. It can therefore also be used in an "ASCII environment" without much hassle.

The GB 2312, which at present is the accepted standard for all kinds of computer applications in China, contains 6,763 simplified *Hanzi*, also divided in Level 1 with 3,755 frequent characters, arranged alphabetically after their reading, and Level 2 with 3,008 rare characters, arranged according to radical and stroke order. A "Secondary Set" of additional characters was repeatedly reported to be in preparation, but has not been published to date.

Other code plans that also contain large numbers of Chinese characters have been brought out in Taiwan (CCCII, a three-byte code that in microcomputer applications has been recently supplanted by the BIG-5 code which again uses two bytes) and Korea (KIPS, Korean Information Processing System).

1.45 Requirements to Recognition Systems

Recognition systems are typically composed of the following sub-systems:

- input
- preprocessing (noise reduction, binarization, segmentation)
- feature extraction
- classification
- output of results.

This structure applies to Latin as well as Chinese character recognition. OCR systems for typewritten Latin letters typically consist of a scanner and

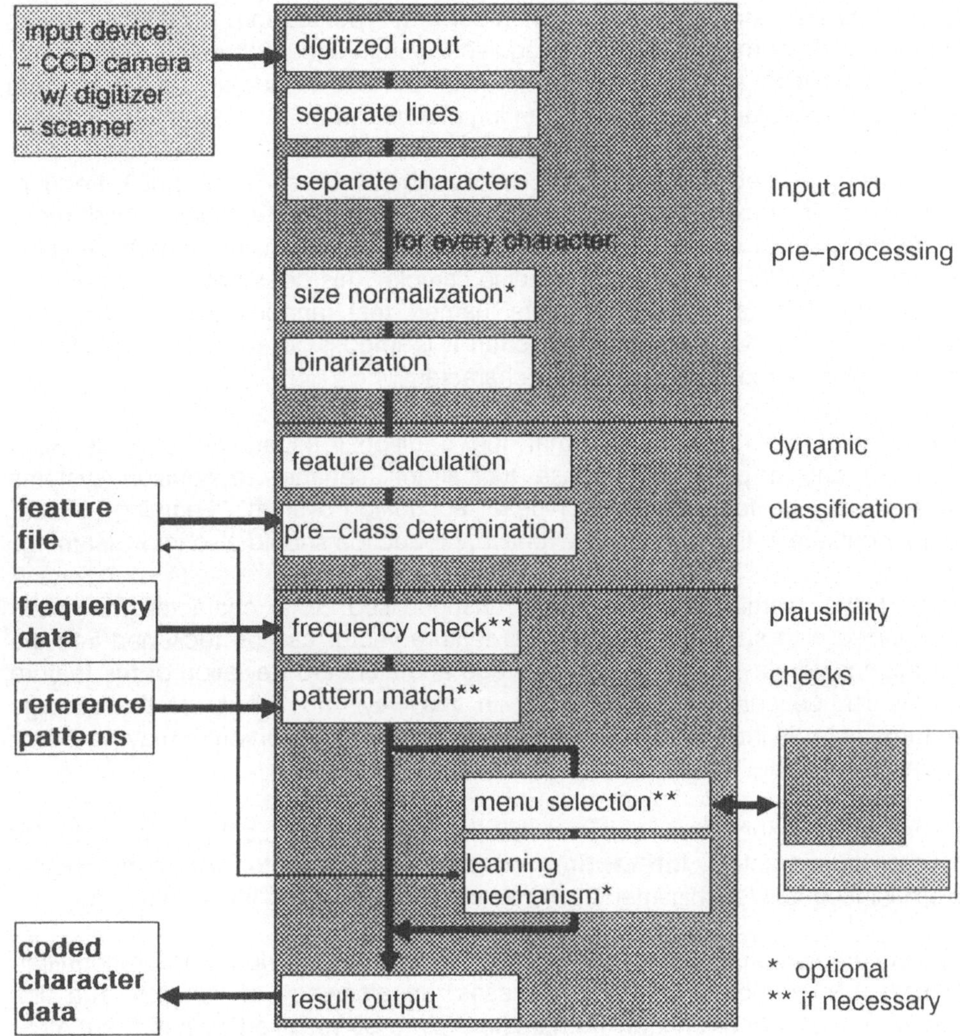

Fig. 1.4-3 Structure of the TECHIS recognition system

recognition software. See [McC 87] for a recent overview of alphanumeric OCR systems that are available in the market. A prototype Kanji OCR system is described in [Hir 80]. Fig. 1.4-3 shows the rough structure of the experimental TECHIS system developed by the authors at the Technical University of Berlin.

If optical recognition of Chinese characters is to be used in real-life applications, the input technology must allow the reading of at least one page, or even better an opening (two facing pages), in one exposition. The fact that Chinese characters are often printed on thin, yellowish dark paper with uneven structure and poor ink must also be taken into account.

Article and chapter headlines are often set in a typeface that is different in size and font style from body text. A recognition system should therefore be possibly **independent of font and size** or at least be able to adapt to different fonts and sizes, for instance by using a learning mechanism.

The character set to be recognized should cover at least 3,000 frequent characters. If necessity arises, it should be easily extendible. Since many characters appear only rarely, the system should be able to produce acceptable recognition results from just one learning sample. This does not conflict with the advantage of larger sample numbers, namely that random noise in individual samples can be reduced, but very often it is impossible to collect a sufficient number of instances of a given, rare, character.

This requirement also implies that the traditional separation of a learning (training) and a production phase typical for alphabet recognition systems cannot really be followed in a Chinese recognition system. "Learning" should rather continue in the "production" phase, production should also imply learning.

Part of the learning can be done unsupervised. If a character has been recognized with sufficient reliability, its feature vector can be reckoned into the corresponding class, and the new average and standard deviation of the feature vector can be computed recursively; in this way, the validity of the average feature vector is increased, and, allowing for possible distortion, it may be better prepared for future instances of the same character class.

In cases of doubt that can not be resolved automatically, especially for out-of-set characters, **interaction** with the human operator, like menu choice, is still indispensable. This interaction can either be done on-line or off-line.

On-line interaction requires the presence of an operator during the recognition process. On the other hand, it increases the efficiency of learning. The first occurrence of a (e.g. unknown) character has to be resolved by hand, but most of the following occurences can already be recognized without need for interaction.

Off-line interaction would reject each of these occurrences, forcing the operator to repeat a decision several times. The advantage of off-line interaction increases with the recognition rate: if in one week of work only a handful of problems is found, the operator can decide on these in few minutes and does not have to wait in the meantime for the recognition to run automatically.

If however the operator is also in charge of feeding the input originals and the speed of recognition is sufficiently high, he might as well keep himself ready for eventually needed interactions.

2 Input and Preprocessing: Setting the Stage

The input and preprocessing procedures create the necessary conditions for the recognition process itself. A number of general image processing methods are applied here in the phases of digitizing (creating a discrete representation of an analog image), segmentation (cutting individual characters out of the image), and picture enhancement. For an introduction to image processing algorithms, see [Daw 87].

2.1 Optical Input

Optical input involves the transformation of pictorial information into digital data, typically a two-dimensional array of picture elements (mostly abbreviated as "pixels" or "pels") that represent the brightness value of tiny segments of the original picture. It is the first step in the recognition of conventional characters on paper, printed or handwritten. Other than optical input methods require special printing techniques, such as magnetic ink, or the presence of the operator to write characters in real time during the recognition process (see 1.43). For the rest of this book, only optical input will be considered.

2.11 Resolution Requirements

The size of a character may either be measured in units of length (millimeters or inches) or, after digitizing, in the number of rows and columns which will be called here the relative resolution. The resolution of an optical input device is given as the ratio between pixels and length units. For scanners, dots per inch (dpi) is the typical measure of absolute resolution.

The smallest feasible size for Chinese characters output on raster devices is about 14*15 pixels (see 1.31). For optical input, higher resolution is required:

"The computer-printed document may have much higher spatial frequencies than the resolution of the output device. For example, a 300-dpi laser printer may produce edges with frequency components of 600 cycles per inch. To scan and reproduce such an image accurately, at least 1200 samples per inch scanning resolution is necessary" [Rub 88:224].

Demands on character recognition can however be lower than on facsimile.

[Suen 86:571] gives suggested matrix sizes of 25*32 to 35*45 for printed and 30*40 to 60*80 for handprinted characters.

This proposal was confirmed by the experiences in the TECHIS project. As the standard character size was fixed at 64*64, an input resolution of about the same values yielded (of course) the best recognition results, but original resolutions down to approx. 30*40 did not notably degrade the recognition quality.

2.12 Camera Input

A video camera transforms a section of a text into analog video signals that represent a video frame, which are then digitized (see below 2.15) into a grayscale picture. While vacuum tube cameras sometimes are still used, CCD (charge-coupled device) cameras deliver better pictures with more even distribution of intensities that need less preprocessing. Text input by video camera has the following properties:

The relative resolution of the digitized text may be varied by modifying the camera-to-text distance. Therefore very small characters can be read. For characters of a given size, various resolutions can be experimented with.

The camera output signal can be fed through to a video monitor, so the user can verify the correct alignment of the text before starting the digitizing process. In this way, rotated characters can be avoided. Character rotation may of course be dealt with by an adequate correlation of picture and text position or by compensating the rotation by software on the digitized image, but these manipulations require considerable computation time. A rotation by a small angle may also impair the segmentation of the image into lines and characters. Finally, the user is free to vary some parameters (illumination, focus, or adjustment of the aperture).

Thus, camera input is useful for experimental research where stable and controlled recording conditions are to be observed.

A drawback of video camera input is that with a typical resolution of 512*512 pixels, only a small section of the original text that is approximately the size of a postage stamp can be read (see fig. 2.1-1 for a typical camera picture). As the width of the input frame is mostly less than the column width of the text, context information between characters or lines is lost. Special handling is necessary to reconstruct the original sequence of characters from the input sequence.

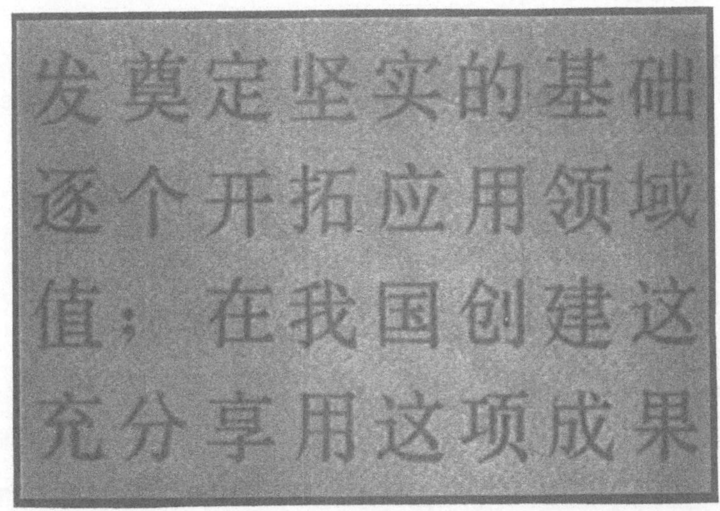

Fig. 2.1-1 A typical camera picture

2.13 Standardizing the Recording Conditions

The quality of camera output images depends on two conditions: illumination and resolution. To guarantee constant illumination and resolution, the following procedure was practiced in the TECHIS project:

A reference pattern (a representation of two grayscales) was recorded by the camera and stored by the computer. Before taking a new shot, the reference image is again read by the camera and the difference to the stored reference pattern is computed dynamically and displayed in real-time. Ideally, if the present illumination and resolution are equal to the stored conditions the difference image should be zero.

In practice, both reference illumination and resolution can only be reproduced approximately and the difference image has to be minimized. Therefore the difference image is displayed on a video monitor. Illumination and resolution are varied until the difference image is darkest.

2.14 Scanner Input

Using a scanner, a page of A4 size (c. 21 by 29 cm) can be read at one time. The resolution can however not be adjusted continuously. At present, scanners with resolutions of 200, 300, 400, 800, 1200 and up to 2000 dpi are available in the market.

These absolute resolutions yield the following relative resolutions (height in pixels; the width is approximately equal) for body text sizes:

Size	300 dpi	400 dpi	600 dpi	800 dpi
8p	33	44	66	88
10.5p	44	58	88	116
14p	58	77	116	155

Scanners deliver a digitized binary or grayscale image. While a grayscale image contains more valuable information, the storage requirements are also much higher than for tightly packed binary images: for 64 gray levels, six times the binary equivalent is required. For digitization and binarization of an A4 page with a resolution of 300 dpi, about 2 Mbyte of data have to be handled. An advantage is that scanners are closed systems with controlled focus and lighting conditions. The quality of the output image will not be influenced by the environment as it happens when reading characters by camera.

2.15 Digitizing Devices

Digitizing devices sample an incoming analog signal, as produced by a video camera, and convert it into a digital signal. For personal computers, relatively small video digitizers are available as peripherals that digitize a video signal with a resolution of 256*256 up to 1024*512 pixels for the whole picture. The digitizing speed varies from near real-time operation (1/25 of a second) to 40 seconds per frame.

For larger computers, picture processing systems are available that perform D/A as well as A/D conversions. With such systems it is possible to digitize analog camera signals, but also to generate video signals for display on an analog monitor.

2.16 Frame Preprocessing

Depending on the actual conditions, the raw picture as delivered from the digitizer or scanner may need preprocessing even before segmentation of the picture into characters. A digitizer may for instance produce high spatial frequencies on horizontal edges, giving them a "half-bitting" appearance after binarization that impairs feature extraction.

In such cases, a simple horizontal filter can be run over the raw frame to equalize this kind of disturbance. Good results were found with the following filter

$$\{f_1(i,j)\} = \begin{bmatrix} 0 & 0 & 0 \\ 0 & 0.5 & 0.5 \\ 0 & 0 & 0 \end{bmatrix} \qquad (2.1-1)$$

that averages each pixel with its neighbor to the right but otherwise has little averaging influence on the image.

Other preprocessing that can be performed at this stage would include filtering of isolated noise pixels, if required by the available input image quality. Since operations on a whole image consume however much more computing time than operations on the small sub-images containing a character each, it seems more sensible to postpone this operation to the single character preprocessing phase. By doing so, the time otherwise needed for filtering noise pixels outside of character cells is also saved.

2.2 Picture Segmentation

After the operations global to the input picture have been completed, the image has to be segmented into sub-images that represent firstly the text lines and secondly the individual characters. To increase overall throughput of the recognition system, the segmentation into lines can also be done in real-time during the scanning process [Lia 87:295], if the hardware permits.

2.21 The Fixed-Distance Approach

A simple approach at segmenting text images expects all characters to be spaced at equal distances both horizontally and vertically. Proceeding from a grayscale image of a section of text, the projection profiles in the directions of row and column of the grayscale image are computed and binarized. The threshold value for binarizing is chosen manually. The interval within which the binarized projection profile has a value of 1 is marked as a possible character.

The sample mean distance of the characters can be computed on the base of the marked co-ordinates for row and column directions. By the results of row and column distances, single characters can be separated from the text.

This procedure has the following disadvantages: Typically in printed, especially Chinese, texts, the distances between characters vary. Moreover the distance characters is supposed to be an integer value. Because the distance is used multiply for the computation of the co-ordinates the quantization error will be summed up. Therefore, segmentation errors will occur frequently.

Instead of the distance between two characters, heighth, width and the specific distance between two neighbouring characters will be used for segmentation from here on. These parameters vary from character to character within a line.

2.22 An Improved Approach

An improvement on the fixed-size method can be made which does not require manual input of the binarization threshold. Instead, a preliminary binarization of the grayscale image takes place in advance.

Let $\{s(i,j)\}$ be a text image with R lines and C columns. In order to compute the starting and ending co-ordinates of each character, the image is first binarized. From the binary image $\{B[s(i,j)]\}$, the horizontal projection profile is computed as

$$P_r(i) = \sum_{j=1}^{C} B[s(i,j)] \qquad i = 1,\ldots,R \qquad (2.2-1)$$

Subsequently, the computed projection profile is normalized within the range of (0..255) according to

$$P_r^* = \frac{255\left[P_r(i) - \min\{P_r(k) \mid k=1,\ldots,C\}\right]}{\max\left\{P_r(k) \mid k=1,\ldots,C\right\} - \min\left\{P_r(k) \mid k=1,\ldots,C\right\}} \tag{2.2-2}$$

and segmented on the base of a threshold. To compensate for single noise pixels, this threshold should be set to a value greater than zero. In the TECHIS system, 5 proved to be a usable value (see fig. 2.2–1).

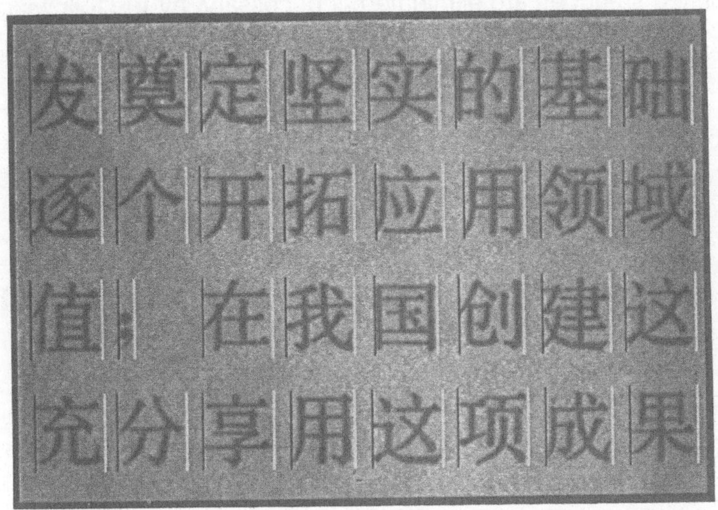

Fig. 2.2–1 Camera picture after segmentation

For exact computation of this threshold, the distribution of the projected disturbances as well as of valid characters has to be estimated. Under sufficient resolution (30 pixel/mm or more) and with text lines parallel to the video scan direction, the segmentation of lines always was correct.

Given a threshold c, the segment of a text line is beginning at the position

$$i = i_b, \text{ if } P_r^*(i_b - 1) < c \leq P_r^*(i_b) \tag{2.2-3}$$

and is ending at the position

$$i = i_e, \text{ if } P_r^*(i_e) \geq c > P_r^*(i_e + 1) \tag{2.2-4}$$

By using these co–ordinates, a single text line

$$\{TL(i,j)\} = \left\{ B[s(i,j)] \,\middle|\, i_b \leq i \leq i_e;\ j=1,\ldots,C \right\} \tag{2.2-5}$$

may be separated from the whole frame and is ready for further computation. Within the image, isolated smudges are eliminated, if necessary, by the following threshold $f_8[.]$:

$$TL^*(i,j) = \begin{cases} 0 & \text{if } f_8\left[TL(i,j)\right] \leq 1 \\ TL(i,j) & \text{if } f_8\left[TL(i,j)\right] > 1 \end{cases} \tag{2.2-6}$$

and the value f_8 [TL(i,j)] is given by the low–pass filtering with eight neighbor pixels of TL(i,j):

$$f_8\left[TL(i,j)\right] = \sum_{n=-1}^{1} \sum_{m=-1}^{1} TL(i+n, j+m) \tag{2.2-7}$$

The vertical projection profile of a text line is computed in analogy to the horizontal one for every text line

$$P_c(j) = \sum_{i=i_b}^{i_e} B\left[s(i,j)\right] \qquad j=1,\ldots,C \tag{2.2-8}$$

This projection profile is normalized as described above and segmented by a threshold with the value 1.

2.23 Check on Correct Segmentation

Some Chinese characters consist of two or more vertically disjunct parts that may result in runs of white columns (gaps) inside the character. Occasionally, such characters will be wrongly separated into two (or even three) parts. Therefore the separated segments have to be checked for correct segmentation.

In many Chinese texts, it can be observed that the space between two neighboring characters may be very wide. Fig. 2.2–2 illustrates the space between two neighboured characters and the space between the segments (gap) of a character.

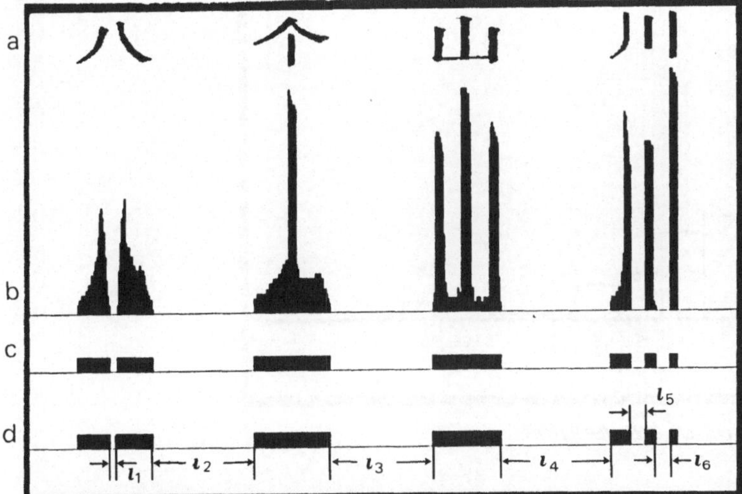

Fig. 2.2–2 Examples for segmentation problems

Fig. 2.2–2(a) shows four Chinese characters. The first character consists of two separated components, while the fourth consists of three separated components. The other characters consist of one component each.

In Fig 2.2–2(b) the normalized vertical projection profile is shown. When this profile is segmented by a threshold value of $\Theta = 1$, segments of different size will be received (fig. 2.2–2(c)). The length of the space between neighboring segments is marked as l_i in Fig. 2.2–2(d).

Comparing Fig 2.2–2(a) with 2.2–2(d), it will be noticed that the set $\Omega_g = \{l_1, l_5, l_6\}$ contains gaps between the segments of one character, while the set $\Omega_s = \{l_2, l_3, l_4\}$ contains spaces between two neighbouring characters.

The relation between the distance of two neighbouring characters and the average width of characters depends on the style of printing and the width of the two adjacent characters, while the relation between the distance of neighbouring segments of a character and the average width of characters depends on the structure of characters as well as on the font.

In fig. 2.2–3, the distribution of the distance in pixels between segments of a character is represented. This statistics was computed for the 3755 reference characters of the font CCS 01–87.

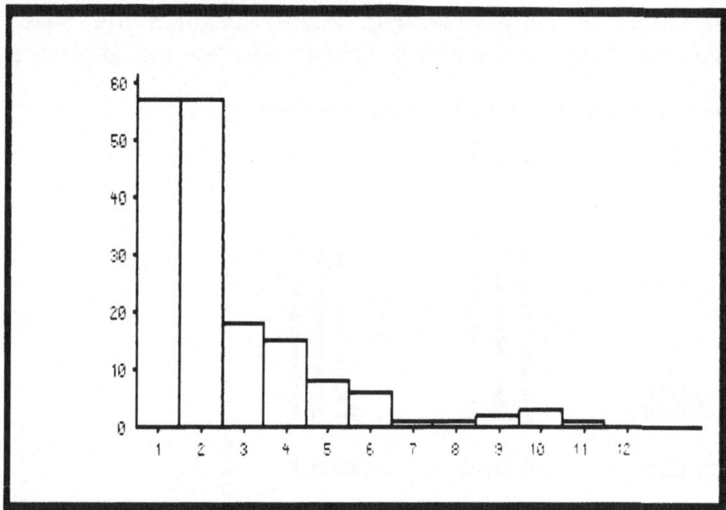

Fig. 2.2–3: Distribution of gap lengths

In this font the maximum width of space between segments of one character equals 11 pixel. The average width of characters of this font runs to 54.3 pixel. The maximum gap width between the segments of one character is about one fifth of the average character width.

If the minimal space between two neighbouring characters in a text line exceeds the maximal gap width between two segments of one character, i.e.

$$\max \left\{ l_i \mid l_i \in \Omega_g \right\} < \min \left\{ l_i \mid l_i \in \Omega_s \right\} \qquad (2.2-9)$$

a decision between two characters or two segments of one character is possible.

If $\Omega_l = \{ l_i \}$ is the set of all gaps of a line, then $`\Omega_l$ can be divided by the threshold value Θ into the classes

$$\Omega_g = \{ l_i | l_i \le \Theta \} \text{ and } \Omega_s = \{ l_i | l_i > \Theta \} ; l_i \in \Omega_l \qquad (2.2-10)$$

and Θ is determined by the condition

$$\max \{ l_i | l_i \in \Omega_g \} < \Theta < \min \{ l_i | l_i \in \Omega_s \} \qquad (2.2-11)$$

If this condition does not hold, the method of regarding the gap lengths leads to wrong segmentation and another technique has to be used.

There is, however, another way to minimize the rate of wrong segmentation. Fig. 2.2-3 shows that gaps inside most of the characters are less than 4 pixel wide. In the font CCS 01-87 there are only 55 characters, or less than 2% of the whole font, in which the gap length between segments is greater than 2 pixels.

If the threshold value Θ of this font is set to 4, the rate of wrong segmentation will be less than 2%. For fixing the optimal value of Θ, the distribution of the gap lengths of the current text has to be determined.

After classification of the gaps Ω_l inside a line, the single characters can be cut from the text.

The algorithm described above was used for reading the reference character font CCS 01-87 (3755 characters). Except for large noise areas, no problems of segmentation were observed.

2.3 Size Transformation

In Chinese texts, different type sizes are employed for headlines, subheads, body text, and footnotes. Size normalization is necessary if the feature algorithm requires input matrices of uniform size. The character matrix as found in the image is changed with scaling factors that have to be determined. Two different approaches to calculating these factors (individual vs. frame-global) as well as three algorithms for performing the actual scaling will be discussed.

Even inside the same font, the dimensions of Chinese characters vary slightly. An analysis of the character width (see fig. 2.3-1) and height (see fig. 2.3-2) of the 3755 characters in CCS 01-87, that were nominally all standardized inside a 64 by 64 pixel window, returned the following results:

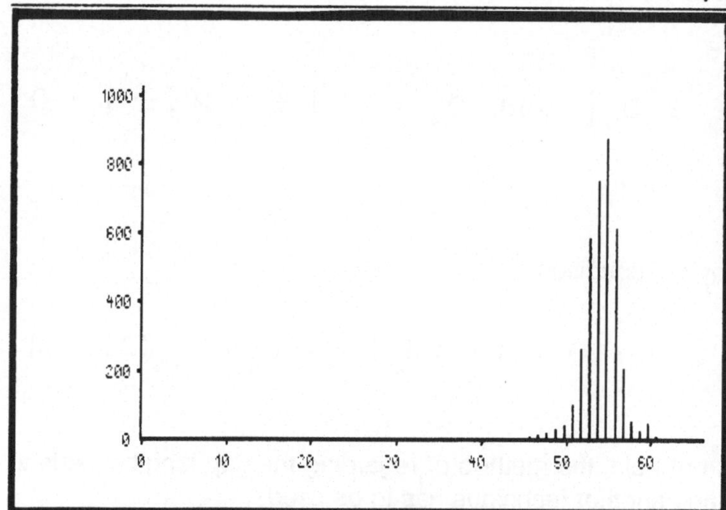

Fig. 2.3–1 Character width distribution

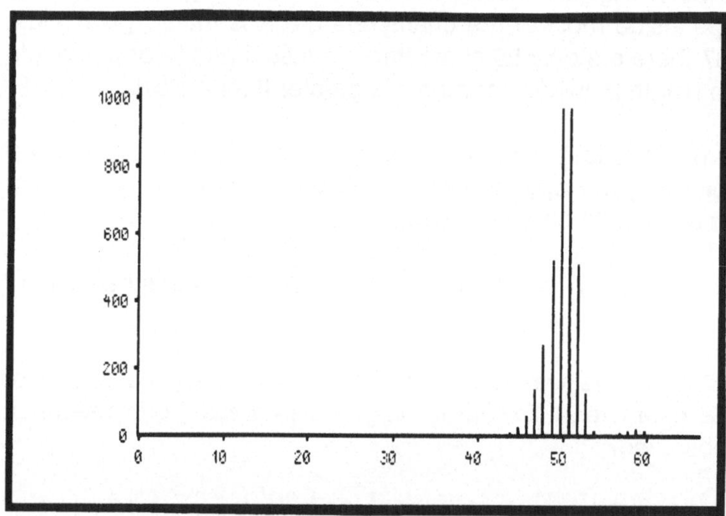

Fig. 2.3–2 Character height distribution

Average character width W_a =	54.3 pixels
Standard deviation	2.4 pixels
Average character height H_a =	50.3 pixels
Standard deviation	2.2 pixels

2.31 Individual Size Determination

If characters of different sizes and fonts occur in one image, it may be necessary to handle every character individually as follows: Let (x_a, y_a) be the co-ordinates of the upper left and (x_e, y_e) be the co-ordinates of the lower right corner as delivered by the segmentation algorithm. The measures (integer numbers) are

$$W = x_e - x_a \quad \text{(width)} \qquad (2.3-1)$$

$$H = y_e - y_a \quad \text{(height)} \qquad (2.3-2)$$

for each character

The scaling factors are computed separately for every character in both directions:

$$k_x(\cdot) = \frac{n_x}{W} \qquad (2.3-3)$$

$$k_y(\cdot) = \frac{n_y}{H} \qquad (2.3-4)$$

where n_x and n_y are the number of rows and columns in the standardized matrix, for instance, $n_x = n_y = 64$ in the TECHIS project. If a field size smaller than the available matrix is used ($n_x = W_a$, $n_y = H_a$), then the character is aligned to the upper and left margins. In this case, it should be centered during the size transformation process.

This individual approach works well with characters of uniform x/y ratio, as most Chinese characters are (see fig. 2.3–1 and 2.3–2). Problems arise when this ratio deviates very much from average. In Chinese texts, narrow punctuation marks are a typical example: the small circle used as a full stop is distorted to elliptical shape.

The distinction between the characters 日 and 曰 is lost. To be able to discriminate them, the dimensions might be stored for later thresholding.

2.32 Frame-global Size Determination

Uniform scaling factors can be determined by using the average width W_a^* and height H_a^* for all characters in the current frame (which of course have to be computed first):

$$k_x^* = \frac{W_a}{W_a^*} \qquad (2.3-5)$$

$$k_y^* = \frac{H_a}{H_a^*} \qquad (2.3-6)$$

The parameter of the standardized matrix can in this case not be chosen as dividend, because otherwise input characters larger than the average would extend beyond the standardized matrix.

This method requires the input characters to be all of uniform size. Care should be taken that headings and body text are not mixed in the same frame. On the other hand, it preserves the proportions of punctuation marks as well.

2.33 Bi-linear Interpolation

A quick size transformation of grayscale images can be done by bi-linear interpolation. Utilizing the scaling factors described above, the intensity is calculated for every pixel of the standardized matrix T = $\{t(i,j)\}$ by examining four pixels of the original matrix S = $\{s(i_s,j_s)\}$. The co-ordinates (i,j) of a given pixel are scaled with the factors as computed above, which normally results in non-integer (virtual) original coordinates:

$$t(i;j) = \sum_{l=0}^{1} \sum_{k=0}^{1} g_{lk}(i;j)\; s(i_s + l; j_s + k) \qquad (2.3-7)$$

The coefficients g_{lk} depend from the (scaling) transformation of coordinates that will now be determined.

$$\tilde{i}_s = \frac{i}{k_x} \qquad (2.3-8)$$

$$\tilde{j}_s = \frac{j}{k_y} \tag{2.3-9}$$

Since only integer co-ordinates can be used in addressing the source matrix, pairs of integer coordinates are computed by rounding:

$$i_s = \text{int}\ (\tilde{i}_s) \tag{2.3-10}$$

$$j_s = \text{int}\ (\tilde{j}_s) \tag{2.3-11}$$

The fractional part of the non-integer co-ordinates determines the relative linear distance to the integer co-ordinates:

$$d_i = (\tilde{i}_s - i_s) \tag{2.3-12}$$

$$d_j = (\tilde{j}_s - j_s) \tag{2.3-13}$$

and it is reasonable to choose theweighting coefficients in the following manner:

$$g_{00}(i;j) = (1-d_i)(1-d_j) \tag{2.3-14}$$

$$g_{01}(i;j) = d_i(1-d_j) \tag{2.3-15}$$

$$g_{10}(i;j) = (1-d_i)\ d_j \tag{2.3-16}$$

$$g_{11}(i;j) = d_i\ d_j \tag{2.3-17}$$

This algorithm performs acceptably fast, but it filters high frequencies because of the averaging. Edges are to a certain degree smoothened, sharp corners are rounded. Our experiments showed, however, that this effect does not notably degrade the recognition rate.

2.34 Cubic Spline Interpolation

Image size transformation can also be done with continuous reconstruction of the image and subsequent loci sampling. This effect can be achieved virtually if the (M x N) original image S = {s(i;j)} is converted into the standardized image T = {t(i;j)} by using cubic spline functions.

First, the number of columns in the standardized image is calculated: Let $\underline{s}(i)$ = {s(i;1),...,s(i,N)} be the i–th row of image S. Using the scaling factor k_x in row direction, the number of columns in the standardized image is given by

$$C = k_x N \qquad (2.3\text{-}18)$$

To convert a sequence $\underline{s}(i)$ into another sequence $\underline{\tilde{s}}(i)$ (one row of the intermediate result), the Hermitic Spline functions are used for interpolation. The elements of $\underline{\tilde{s}}(i)$ are computed [Ros 82] as

$$\underline{\tilde{s}}(i;\mu) = \underline{\alpha}_\mu^T \, \underline{Q} \, \underline{c}_\mu(i) \quad i=1,\dots,M; \; \mu=1,\dots,C \qquad (2.3\text{-}19)$$

$$\text{with} \quad \alpha_\mu = \left(\frac{\mu}{k_x} - \left[\frac{\mu}{k_x} \right] \right),$$

$$\underline{\alpha}_\mu = [\, \alpha_\mu^3, \alpha_\mu^2, \alpha_\mu, 1 \,],$$

$$\underline{Q} = \frac{1}{6} \begin{bmatrix} -1 & 3 & -3 & 1 \\ 3 & -6 & 3 & 0 \\ -3 & 0 & 3 & 0 \\ 1 & 4 & 1 & 0 \end{bmatrix}$$

and

$$\underline{c}_\mu(i) = \Big[s(i;\mu-1), s(i;\mu), s(i;\mu+1), s(i;\mu+2) \Big]^T .$$

Subsequently, the columns of the (M x C) intermediate matrix $\tilde{S}$

$$\underline{\tilde{s}}(j) = [\, \tilde{s}(1;j), \dots, \tilde{s}(M;j) \,]^T \quad j=1,\dots,C \qquad (2.3\text{-}20)$$

are transformed with the same method to columns of the (R x C) standardized matrix T = {t(i;j)} regarding the scaling factor k_y in column direction, i.e.

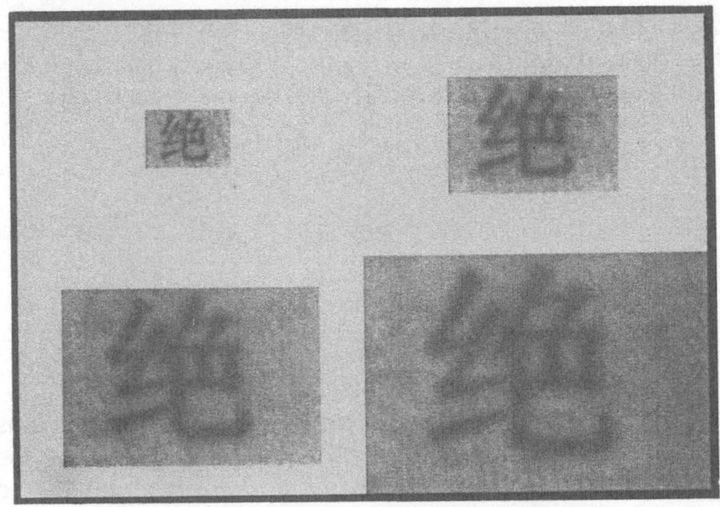

Fig. 2.3–3 Size normalization of grayscale pattern

$$R = k_y \ M, \tag{2.3-21}$$

$$\underline{t}(j) = [t(1;j), \ldots, t(R;j)]^T \quad j = 1, \ldots, C \tag{2.3-22}$$

with

$$t(\mu;j) = \underline{\tilde{\alpha}}_\mu \ \underline{Q} \ \underline{\tilde{c}}_\mu(j).$$

This method results in better standardized images (see fig. 2.3–3), but requires considerably more time for computation.

2.35 Size Normalization on Binary Patterns

The normalization methods discussed so far work best with gray–level images. Sometimes only binary patterns are available for recognition, for example from a scanner that delivers binary pictures. For this case and for features that demand complete alignment of the character in the matrix, a special normalization algorithm that also filters isolated noise spots from the background was developed.

Horizontal and vertical projection profiles (see 3.23) can be separated into several parts. If a segment of the projection is smaller than a certain threshold, it is supposed to be spot and ignored further on. Consequently, a misextraction of the rectangular frame is avoided.

The pixel coordinates of the rectangular character frame are represented by i_1, i_2, j_1 and j_2. Through normalization, the image within the frame is mapped into a, for example, 64*64 matrix. Two simple mapping methods have been tested in experiment. In the first method, the rectangular frame in the original image F = {f(m;n)} is mapped into a square frame H = {h(i;j)} of 64*64 pixels as follows:

$$h(i;j) = f\left[\operatorname{int}\{i_1 + k_x(i-1)\};\ \operatorname{int}\{j_1 + k_y(j-1)\}\right]$$

$$i,j = 1,\dots,64 \qquad (2.3-23)$$

with

$$k_x := \frac{(i_2 - i_1)}{63}, \quad k_y := \frac{(j_2 - j_1)}{63}.$$

In the second method, the mapping factors are equal in both directions, thus preserving the aspect ratio of the character. If the height is greater than the width, the factor should be determined by the height. In horizontal direction, a shift should be made so that the character can be centered in the standardized matrix.

The mapping formula of the second method is shown as follows:

Calculate the parameters

$$k_x := \frac{(i_2 - i_1)}{63}, \quad c_o := \frac{(j_2 - j_1)}{(i_2 - i_1)}. \qquad (2.3-24)$$

$$L := 1 + c_o\, 63 \qquad (2.3-25)$$

$$c_1 := 1 + \frac{(64 - L)}{2} \qquad (2.3-26)$$

$$c_2 := \frac{(64 + L)}{2} \qquad (2.3-27)$$

The transformed image is then given by

$$h(i;j) = \begin{cases} 0, & \text{if } j < c_1 \text{ or } j > c_2 \\ f\left[\operatorname{int}\{i_1 + k_x(i-1)\};\ \operatorname{int}\{j_1 + k_x(j-c_1)\}\right] & \text{otherwise} \end{cases}$$

$$i,j = 1,\dots,64 \qquad (2.3-28)$$

The second mapping method can preserve the aspect ratio of character, but this is not always an advantage. For example, if two fonts differ from each other mainly on the aspect ratio, the first mapping method neutralizes this difference and thus is more suitable.

In general, the ratios of horizontal and vertical resolution are not identical for different kinds of input devices. If the reference samples have been read by video camera and the character samples to be recognized are read by scanner, characters of the two sets may have different ratio of height and width. If the first mapping method is adopted, the aspect ratio is unified for all characters, thus a good match between input and reference pattern can be achieved. Compared to the second mapping method, the first method is more suitable to recognition, though less in accordance with human perception.

These normalization methods are very simple and do not require much calculation, but the quality of the normalization is only intermediate. Some deformation can be caused to slanting strokes, especially when the character size is much less than 64*64, because the size normalization is achieved simply by duplicating or deleting some lines and columns (see fig. 2.3–4).

Fig. 2.3–4 Size normalization of binary patterns

Another method for size normalization of binary patterns is to treat the binary image as a grayscale pattern (with gray–levels 0 and 1), then to perform cubic spline interpolation (see 2.34), and finally to re–binarize the resulting picture with a fixed threshold of 0.5. The geometrical features of the character are better preserved (see fig. 2.3–5), but the computational load is higher as well.

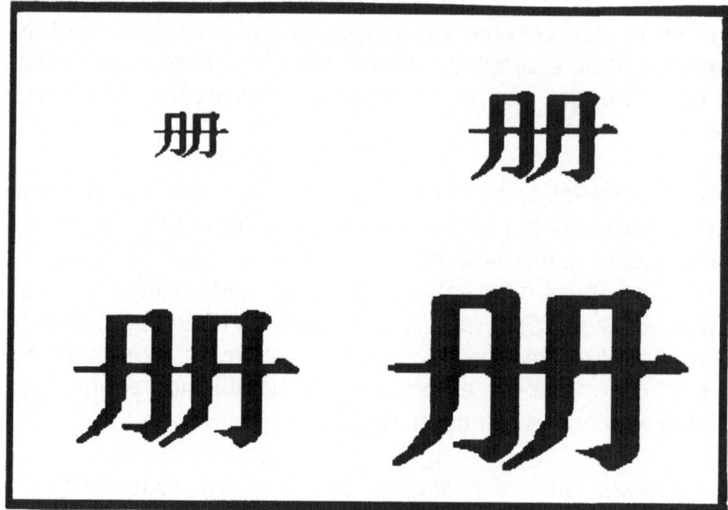

Fig. 2.3–5 Spline interpolation on binary patterns

2.4 Binarization

Binarization is the process of converting a grayscale image into a binary image whose pixels may only be white or black. The advantage of binary images is that they can be stored in much less space and thus simplify the feature calculations. On the other hand, a certain amount of information is lost. Since original printed characters normally are bimodal (black on white) anyway, the binarization reverts the effect produced by the camera: "Indeed, conventional television cameras are designed to provide gray transitions around sharp boundaries by averaging the image" [Rub 88: 111].

This sequence of averaging and re-thresholding carries the risk of information losses. For instance, thin strokes may disappear. On the other hand, a careful selection of the binarization threshold allows stroke thinning, leading to clearer binary images (see fig. 2.4–1).

For binarizing characters, several methods can be employed, the simplest being a global threshold algorithm. Discriminate analysis [Ots 79] of the distribution of grayscales produces a threshold valid for the whole image:

Let S = {s(i,j)} be a grayscale image. Create a histogram (relative frequency)

$$p_s(g) \;:=\; \frac{n_g}{G}, \quad g = 0, \ldots, (G-1) \qquad (2.4-1)$$

where

n_g := number of values with graylevel g
G := total number of graylevels

This amounts to saying

$$\sum_{g=0}^{(G-1)} p_s(g) = 1 \tag{2.4-2}$$

The relative frequency of the sum is defined as

$$h_s(g) := \sum_{k=0}^{g} p_s(k) \tag{2.4-3}$$

The mean value in the histogram is given by

$$m_s := \sum_{k=0}^{(G-1)} k\, p_s(k) \tag{2.4-4}$$

and the mean value for truncated graylevel at g is

$$m_s(g) := \sum_{k=0}^{g} k\, p_s(k) \tag{2.4-5}$$

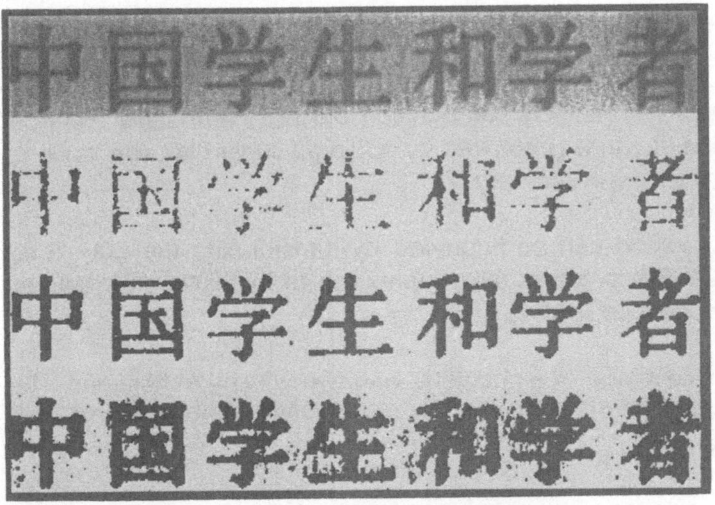

Fig. 2.4–1 Binarization with various thresholds

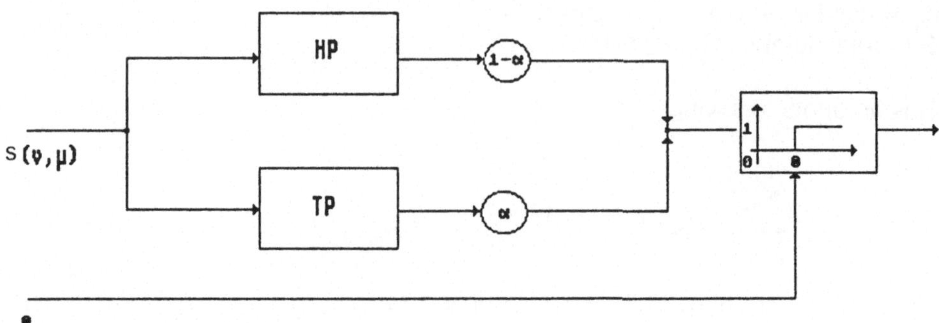

Fig. 2.4–2 Principle of binarization

To determine a suitable threshold of S, it makes sense to find the maximum weighed square deviation of mean values $m_S(g)$ in dependence of the gray level g, i.e.

$$f(g^*) = \max_g \left(\frac{h_g(g)}{1-h_s(g)} [m_s(g)-m_s]^2 \right) \qquad (2.4-6)$$

and the binarization threshold can be determined as

$$\Theta_s = g^* - 1 \; . \qquad (2.4-7)$$

This threshold is usable for character recognition but not necessarily identical with human expectation. It can be improved by adding a linear bias, the value of which must be determined experimentally.

The global threshold method can be improved by transforming the gray–level picture (see fig. 2.4–2). This prevents the suppression of thin lines and reduces the influence of inhomogenous lighting.

Let s(i,j) be the grayscale image of a character with N rows and M columns. The edges in this image are first enhanced by an appropriate combination of high pass and low pass filters:

$$u(i,j) = \left[\alpha s_T(i,j) + (1-\alpha) s_H(i,j)\right], \quad 0 \le \alpha \le 1 \qquad (2.4-8)$$

A low pass (average) filter is chosen:

$$s_T(i,j) = \frac{1}{5}\left[s(i-1,j)+s(i+1,j)+s(i,j)+s(i,j-1)+s(i,j+1)\right] \quad (2.4-9)$$

A simple high pass filter is defined by the difference between original and low pass image (**Laplace filter**):

$$s_H(i,j) = s(i,j) - s_T(i,j) \quad (2.4-10)$$

Finally, the enhanced image is binarized with a fixed threshold θ_S, yielding the binary image B = {b(i;j}:

$$b(i,j) = \begin{cases} 1 & \text{if } u(i,j) \geq \Theta_s \\ 0 & \text{if } u(i,j) < \Theta_s \end{cases} \quad (2.4-11)$$

2.5 Edge Smoothing on Binary Patterns

In the binary image of printed Chinese characters, there are often a lot of defects on some edges of strokes, especially the horizontal and vertical strokes (see examples in fig. 2.5–1). This kind of defect heavily degrades the results of some recognition features (e.g. the SDF feature), making them rather unstable. When such features are used, preprocessing must include edge smoothing.

Traditional algorithms of image smoothing are not best suited for this particular problem, because this kind of defect can be detected only by analyzing a whole section of the edge. The following improved method can smooth out horizontal and vertical edges effectively, while having little effect on sloping edges.

The image is scanned in up, down, left and right directions. During the scanning in a given direction, the edges paralleling it can be smoothed out. Taking the upside edges as example, the algorithm can be explained as follows.

The image to be processed is represented as:

$$b(i,j) = \begin{cases} 1 & \text{for black pixels} \\ 0 & \text{for white pixels} \end{cases} \qquad \begin{matrix} i=1,m \\ j=1,n \end{matrix} \quad (2.5-1)$$

The pixels lying on upside edges and satisfying the following condition are elements of a set E which is defined as:

$$E = \{ b(i,j) \mid b(i-1,j)=0 \wedge b(i+1,j)=1 \} \qquad (2.5-2)$$

While scanning along the i–th line of the image, it is possible to detect a sequence of upside edge pixels which should be followed as long as possible. The sequence is a whole section of an upside stroke edge and is represented as a set:

$$P_i(l_1, l_2) = \left\{ b(i,j) \mid b(i,j) \in E,\ l_1 \leq j \leq l_2 \right\} \qquad (2.5-3)$$

The sequence P can be divided into several sub–sequences. A sub– sequence is composed by pixels of a same value, 0 or 1. Any two neighbouring sub–sequences must have different values. A criterion has been found to detect the defective edge:

$$(N_s > 3) \text{ or } (N_s > 1) \text{ and } \min \left[\frac{W}{(W+B)} ; \frac{B}{(W+B)} \right] < \Theta \qquad (2.5-4)$$

where N_S is the number of sub–sequences, W is the number of white pixels, B is the number of black pixels and θ is a threshold.

When a defective edge is detected, it is smoothed out. If the number of white pixels is less than the number of black pixels, all white pixels are replaced with black pixels; otherwise black pixels are replaced with white pixels.

Fig. 2.5–1 Edge smoothing

3 Feature Extraction

3.1 Principles of Feature Extraction

Feature extraction is one of the crucial stages in all pattern recognition systems. Extracting features is as important as the problem of making optimum decisions in the subsequent classification stage (see chapter 4 below).

In character recognition, such features have to be found and extracted that are able to separate one class of characters from the others in the feature space. Transformations must (and can) be found that map the normalized characters into a smaller set of features that still contain all the relevant information needed for automatic recognition. These sets of information are denoted as points in the feature space, which is a n-dimensional normed space.

To sum up, features have to describe the given pattern as detailed as required and to yield sufficient separability between character classes. Care should also be taken that the selected features are stable, i.e. that their variation between different samples of the same class is minimal or, in other words, the conditional density function related to the class can be determined.

The selection of appropriate features is the most important problem in character recognition. Various features have been proposed and tested so far. Neverthe-less, it is seen that precise recognition is difficult to achieve when using any single feature except pattern match (which has other disadvantages). This is because any feature lays particular emphasis on one certain aspect and ignores all of the other aspects of characters. It is very well conceivable that some different characters yield an equal or very similar feature value and thereby misrecognition can be caused. A combination of several features can be used to further increase the recognition rate, as will be shown later in this chapter.

The distinctive features used in character recognition can be divided in global and structural methods. Global features, as for instance transformational procedures, counting along certain directions or positions of points and their distances from a reference point, are less sensitive against noise and minor local distortions. Structural features, for example edges and lines, outlines or centerlines of characters, describe the topology of a character more accurately, they are less sensitive to deformations of the character than global features. A feasible character recognition system therefore has to use a combination of methods, each of which may compensate for the shortcomings of the other.

3.2 Useful Tools

The operations described in this section do not deliver values that can directly be used as features. As they are employed in the process of computing several features, though, we will describe these "tools" separately to maintain clarity.

3.21 Blackness

In a (m x n) binary character image B={b(i;j)}, the number of black pixels (the zeroth moment) is easily computed by

$$m_{oo} = \sum_{i=1}^{m} \sum_{j=1}^{n} b(i;j) \qquad (3.2-1)$$

This value is not yet a useful feature, since it depends strongly on parameters such as paper and ink color, lighting conditions, and also strongly on the binarization threshold. It may however be put to good use in conjunction with other features (see below).

If the pixels of an (m x n) image matrix are arranged into a vector $\underline{v} \in \mathbb{R}^{nm}$, the number of black pixels is likewise determined by adding the pixel values:

$$m_{oo} = \sum_{k=1}^{nm} v(k) \qquad (3.2-2)$$

The number of black pixels (the absolute blackness) may be divided by the total number of pixels to indicate the "relative blackness" of the character, but this value is likewise unusable for recognition.

3.22 Stroke Width

Blackness is very sensitive to variations of stroke width, which is in turn influenced by font, printing quality and the process of binarization. In fact, blackness varies as a linear function of the stroke width. A stable feature, Total Stroke Length (see below), can thus be derived from the ratio of blackness and stroke width if the latter can be measured.

The average stroke width can be estimated by the following method [Bar 68]. In a binary image of a character, it is very easy to count the number of positions on which the following template is entirely included in the image:

bb
bb

where the letter "b" represents a black pixel. Assuming that the character under consideration is composed only of one ideal rectangular stroke represented by W rows and L columns, the total number of positions which meet the above condition is determined by the stroke length L and the stroke width W:

$$Q = (L-1)(W-1) \qquad (3.2-3)$$

If L>1, an approximate representation of Q can be made:

$$\hat{Q} = L(W-1) = m_{00}\frac{W-1}{W} \qquad (3.2-4)$$

where m_{00} = LW is the blackness. With the above formula, the stroke width can be roughly estimated as

$$\hat{W} = \frac{m_{00}}{(m_{00} - \hat{Q})} \qquad (3.2-5)$$

Because Chinese characters are composed of slanting as well as horizontal and vertical strokes, the value of Q is actually calculated according to the number of positions on which any of the following three templates is entirely included in the image:

```
  bb  bb   bb
bb    bb    bb
```

This method was originally developed with the purpose of regulating the threshold of binarization dynamically to avoid strong variations of stroke width. The method is adopted here to overcome the shortcoming of the blackness and obtain a more stable rough feature (see below, Total Stroke Length).

3.23 Projection Profiles

Projection profiles have been discussed in the literature as early as 1972 [Nak 72]. For a (m x n) matrix, the pixel values are added separately for the scan lines in each projection direction (typically horizontally and vertically, sometimes also 45^0 and 225^0). The fact that pixel values are added implies that projection profiles are not limited to binary patterns, but may also be used on gray-level matrices.

The resulting vectors that may also be visualized as curves, see fig. 3.2-1, still need to be post-processed. The simplest feature that can be derived from projection profiles is the number of strokes parallel to the scan lines. This involves determining a suitable threshold and counting the number of peaks above this threshold. In a slightly more sophisticated version, the height of the peaks, either raw in pixels or scaled to the size of the matrix or character, is summed up. Both these features are applicable only for characters of limited complexity. A higher number of strokes tends to "flatten out" the projection profiles.

Other approaches included elastic matching of projection profiles by Fourier transform. Although stroke thinning was applied to neutralize stroke width, an error rate of 17 % at a search time of 50 seconds/character was reported [Nak 73].

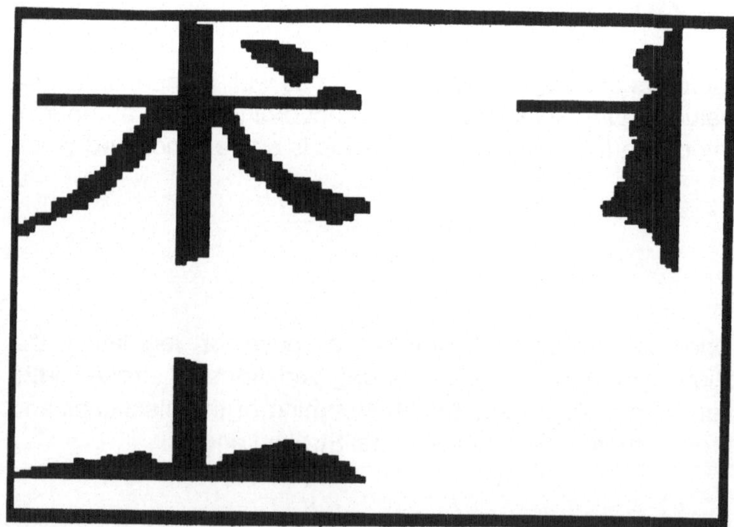

Fig. 3.2-1 Projection profiles

3.24 Transitions

A transition is defined as a significant change of value that occurs while scanning a sequence of matrix (or vector) elements in a given direction (a pixel position may cause a transition in one direction but not in another). In gray–level patterns, the significance of a change has to be determined by a threshold value that may be global (constant for the whole pattern) or local (variant depending on pattern context). In binary patterns, only transitions from 0 to 1 (the so called "**black jumps**") and from 1 to 0 ("white jumps") are possible. Since the numbers of black and white jumps in a given scan run differ maximally by 1, it is sufficient to count only one kind, usually the black jumps.

The number of black jumps in a vector $v \in \mathbb{R}^{mn}$ is determined by

$$N_j(\underline{v}) = \sum_{k=2}^{N} \sigma[v(k) - v(k-1)] \qquad (3.2-6)$$

where the unit step function $\sigma[.]$ is defined as

$$\sigma[x] = \begin{cases} 1, & \text{if } x > 0 \\ 0, & \text{if } x \leq 0 \end{cases} \qquad (3.2-7)$$

Note that transitions can be used to localize strokes independent of their width. In some cases, black jumps in horizontal and vertical direction can substitute the character skeleton which requires considerably more computation effort (see fig. 3.2–2 for an example).

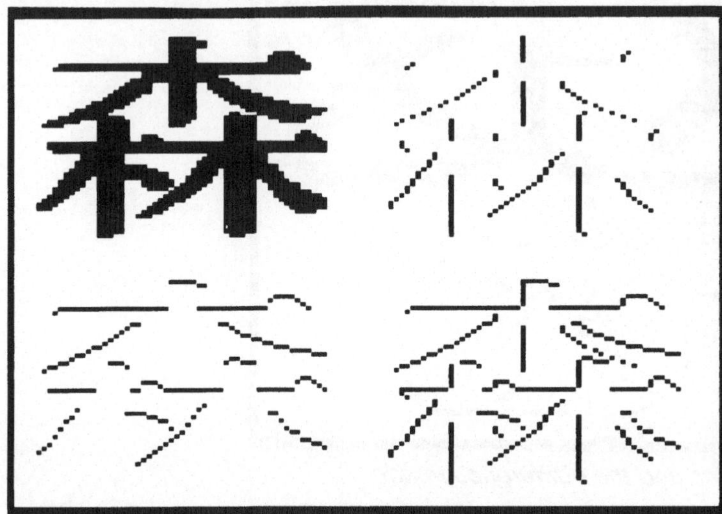

Fig. 3.2–2 Transitions in +x and –y directions

3.3 Some Feature Algorithms

3.31 Pattern Match

The most straightforward approach to character recognition is pattern or template match where two patterns are compared, typically with the pattern distance feature. This method requires the existence of an input matrix as well as of a reference matrix of the same dimensions, often called template, for every character class. If the input matrix is of different dimensions, size normalization is indispensable.

Let $B=\{b(i;j)\}$ be the (m x n) input pattern and $R_x=\{r_x(i;j)\}$ the reference pattern for character class x. Then a possible **pattern distance**, the Hamming distance, is computed as

$$\delta(B, R_x) = \sum_{i=1}^{m} \sum_{j=1}^{n} |b(i,j) - r_x(i,j)| \qquad (3.3-1)$$

In fig. 3.3–1, two instances of the same character with the resulting difference image are shown. The Hamming distance may be considered as the number of black pixels in the difference image.

Fig. 3.3–1 Two patterns and their difference image

A size-independent pattern distance may be given as

$$\delta'(B, R_x) = \frac{\delta(B, R_x)}{m\,n} \qquad (3.3-2)$$

The pattern distance can be used as a feature to select the class with the least distance. Since the feature value is, nevertheless, dependent on both the input and the current reference pattern, the pattern distance has to be calculated once for every class.

Hence, the most important drawback of the pattern match feature is the fact that for N recognition classes, the reference patterns have to be loaded from mass storage and the pattern distance has to be computed N times. The enormous time requirement prevents its straightforward application in the early stages of Chinese character recognition. Nonetheless, pattern match may successfully be employed for discrimination between a small number of candidates, as done in the TECHIS system.

Note that the term "pattern matching" is also used by some authors, e.g. [Mae 82], for different features described hereafter, like the peripheral feature.

A feature derived from pattern match is the **mesh feature** [Ume 82] in which the binary image is divided into a number of subfields (e.g. 8*8) of equal size. The pixel values in each mesh area are summed up, resulting in 64 feature values. This feature produced a recognition rate of 91.48 % and an accumulated classification rate at the tenth order (i.e. inclusion of the correct result in the top 10 candidates) of 99.80 %.

3.32 Peripheral Features

The outlines of a character pattern can be described by **peripheral features** [HUM 79, Ume 82], where the distances between the matrix edge and the first (ET1) resp. second (ET2) black jump in a number of segments (typically eight) are summed.

The extraction method of the ET1 feature is shown in fig. 3.3-2. The character pattern is divided into eight horizontal stripes and eight vertical stripes. Within each stripe, the area between the edge and the first change from white pixel to black pixel (black jump) is calculated. This operation generates a 32-dimensional feature vector (4 edges by 8 stripes).

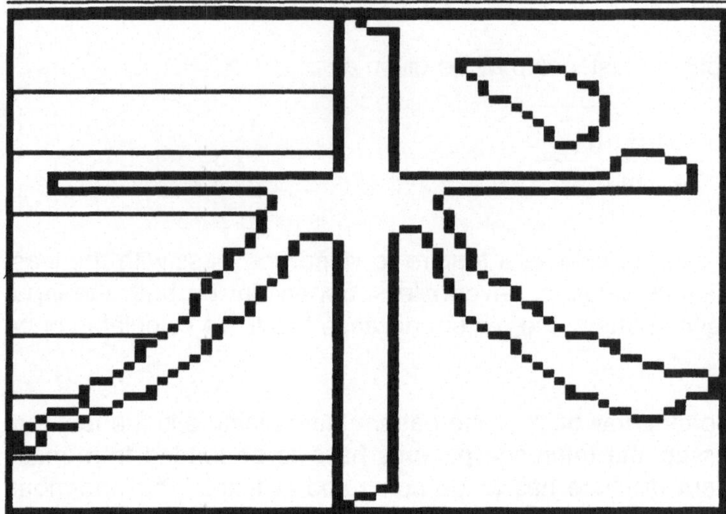

Fig. 3.3–2 Principle of the ET1 feature

For a (64 x 64) binary image matrix B={b(i;j)}, the ET1 feature is defined as

$$ET1(0,n) = \sum_{i=1}^{8} \min\{j \mid b(8n+i;j)=1\} \qquad (3.3-3)$$

$$ET1(1,n) = \sum_{i=1}^{8} \left[64-\max\{j \mid b(8n+i;j)=1\}\right] \qquad (3.3-4)$$

$$ET1(2,n) = \sum_{j=1}^{8} \min\{i \mid b(i;8n+j)=1\} \qquad (3.3-5)$$

$$ET1(3,n) = \sum_{j=1}^{8} \left[64-\max\{i \mid b(i;8n+j)=1\}\right] \qquad (3.3-6)$$

$$n=0,\ldots,7$$

The classification result of ET1 is shown in fig. 3.3–3, where the abscissa shows the order of the candidate categories and the ordinate shows the accumulated classification rate. Categories are ranked in order of distance. The **accumulated classification rate** at the k–th order is defined as the percentage ratio where the correct category is contained in the candidate categories above the k–th order. The accumulated classification rate at the first order is the so–called **recognition rate**.

Feature ET2 can represent the arrangement of strokes inside the character. The character pattern is also divided into eight horizontal and eight vertical stripes. The area between the edge and the second change from white pixel to black pixel is calculated within each stripe so that again a (4 x 8) dimensional vector is

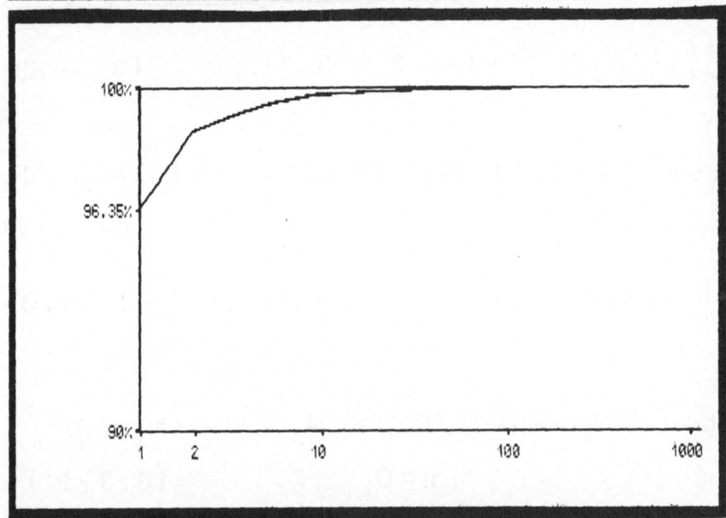

Fig. 3.3–3 Classification results of the ET1 feature

produced. Fig. 3.3–4 illustrates the extraction method of these features. A fixed number c of black jumps in a row of a binary image matrix is determined by

$$N_J(i;j^*) = \sum_{j=1}^{j^*} \sigma\left[b(i;j) - b(i;j-1)\right] = c \qquad (3.3\text{-}7)$$

with $\sigma[.]$ as unit step function (see 3.24). The number of black jumps in a column is analogously defined, i.e. $N_J(i^*;j)$.

For a (64 x 64) binary image matrix B={b(i;j)}, the feature ET2 is defined as:

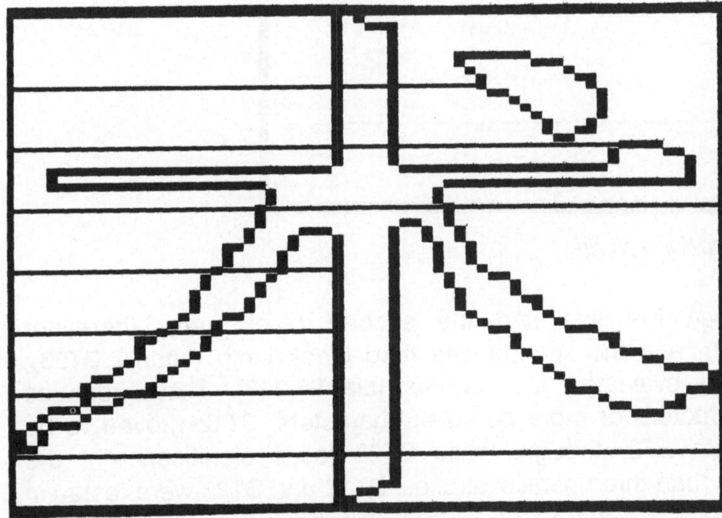

Fig. 3.3–4 Principle of the ET2 feature

$$ET2(0,n) = \sum_{i=1}^{8} \{j^* \mid N_J(i;j^*-1) < 2 = N_J(i;j^*)\} \quad (3.3\text{-}8)$$

$$ET2(1,n) = \sum_{i=1}^{8} \left(64 - \{j^* \mid N_J(i;64) - N_J(i;j^*) = 1;\ N_J(i;j^*-1)\} < N_J(i;j^*)\}\right) \quad (3.3\text{-}9)$$

$$ET2(2,n) = \sum_{j=1}^{8} \{i^* \mid N_J(i^*-1;j) < 2 = N_J(i^*;j)\} \quad (3.3\text{-}10)$$

$$ET2(3,n) = \sum_{j=1}^{8} \left(64 - \{i^* \mid N_J(64;j) - N_J(i^*;j) = 1;\ N_J(i^*-1;j) < N_J(i^*;j)\}\right)$$

$$n = 0, \ldots, 7 \quad (3.3\text{-}11)$$

Experiments conducted on the CTX1 characters (see 5.42) yielded surprisingly high recognition rates for peripheral features (96.35 % for ET1, 98.90 % for ET2). The classification results are visualized in figs. 3.3–3 and 3.3–5. Under better conditions with an (80 x 80) resolution, Umeda [Ume 82] reported even 99.80%.

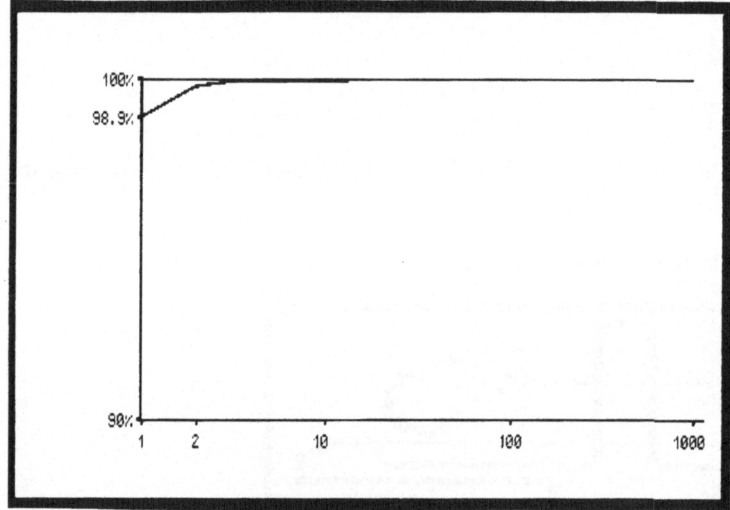

Fig. 3.3–5 ET2 classification results

The distances between the first and the second black jump (hereafter abbreviated with DT12), and the second and third black jump (shortly DT23), respectively, were also evaluated for features [Lia 87:296]. These features express the internal structure of more complex characters. DT12 proved to be strongly correlated with ET2, though, while DT23 loses significance if the character contains less than three parallel strokes. DT12 and DT23 were tested in combination with the 4–SDF feature (see below), but thereafter discarded from further research.

3.33 Stroke Density

The 4–SDF (stroke density function in four directions) proposed by Hagita/Umeda/Masuda [HUM 79] for handwritten Chinese character recognition can be used to represent the stroke density on the different regions.

The (64 x 64) character pattern B={b(i;j)} is evenly divided into eight regions in the directions with the angle of 0, 45, 90 and 135 degrees respectively. In each direction, the numbers of transitions from white pixel to black pixel (black jumps) are summed up within each region. Fig. 3.3–6 is a illustrative example of this feature.

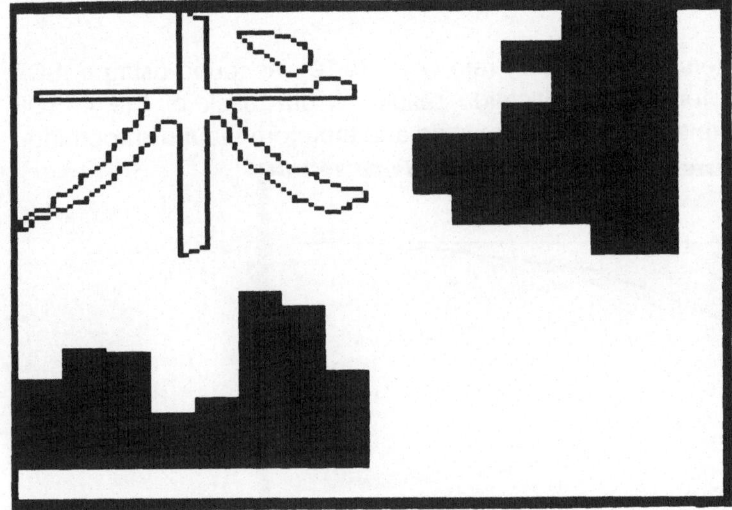

Fig. 3.3–6 Principle of the 4–SDF feature

The 4–SDF feature, here abbreviated as F4, is defined as follows (for notations, see section 3.32):

$$F4(0,n) = \sum_{i=8n+1}^{8n+8} N_J(i;64) \qquad (3.3\text{-}12)$$

$$F4(1,n) = \sum_{j=1}^{64} \sum_{i=8n+1}^{8n+8} \sigma\left[b(i+1;j) - b(i;j)\right] \qquad (3.3\text{-}13)$$

$$F4(2,n) = \sum_{j=1}^{63} \sum_{i=8n+2}^{8n+8} \sigma\left[b(i-1;j+1) - b(i;j)\right] \quad (3.3-14)$$

$$F4(3,n) = \sum_{j=1}^{64} \sum_{i=8n+2}^{64} \sigma\left[b(i-1;j-1) - b(i;j)\right] \quad (3.3-15)$$

$$n = 0, \ldots, 7$$

For the CTX1 characters, a recognition rate of 94.49 % was found using 4–SDF only (see fig. 3.3–7 for the classification result). A drawback of the 4–SDF features is that they are sensitive to edge noise and therefore require smoothing.

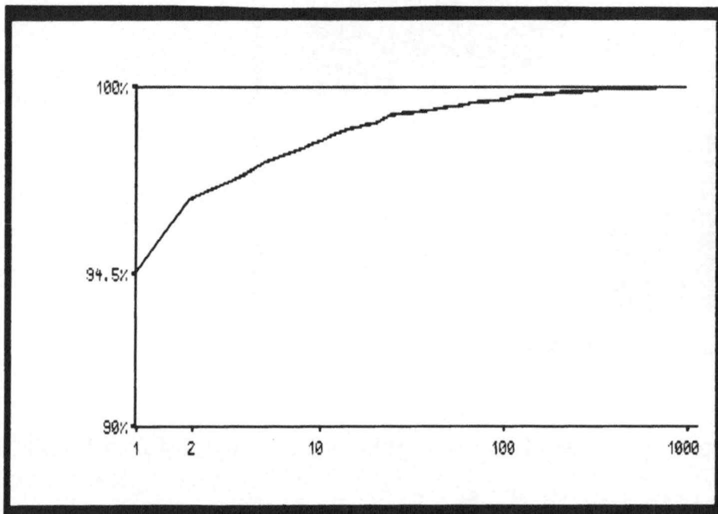

Fig. 3.3–7 Classification results of 4–SDF

3.34 Local Direction Contributivity

The Local Direction Contributivity (LDC) feature was also employed by Hagita/Masuda [Hag 81] in handwritten characters recognition. For every black pixel in the pattern, the bidirectional run length in the four standard directions was determined. These run lengths were added for every direction, resulting in four feature values.

For every pixel in a stroke of given direction, the contribution to the feature value of that direction reflects the stroke length, while the remaining three contributions depend mostly on stroke width. In handwritten characters, especially when done in pen or pencil, an uniform stroke width may be expected. On the other hand, printed fonts differ widely in stroke width. This makes LDC values of printed characters rather font-dependent.

From a certain black pixel (i,j), we can measure the distances to the contour of the stroke in 8 different directions shown in fig. 3.3-8. By summing the distances in the opposite direction together, a local feature vector with 4 components is obtained:

$$d_0(i,j) = \min\{k \mid b(i,j-k)=0\} + \min\{k \mid b(i,j+k)=0\} - 1 \tag{3.3-16}$$

$$d_1(i,j) = \min\{k \mid b(i-k,\ j+k)=0\} + \min\{k \mid b(i+k,j-k)=0\} - 1 \tag{3.3-17}$$

$$d_2(i,j) = \min\{k \mid b(i-k,j)=0\} + \min\{k \mid b(i+k,j)=0\} - 1 \tag{3.3-18}$$

$$d_3(i,j) = \min\{k \mid b(i-k,j-k)=0\} + \min\{k \mid b(i+k,j+k)=0\} - 1 \tag{3.3-19}$$

These local features can determine the direction of the stroke on which the black pixel is located. For example, if $d_2(i,j)$ is much larger than the other three values, the pixel (i,j) lies certainly on a vertical stroke.

On the basis of these local features, the LDC feature is calculated in the following way. The local features are normalized by

$$d_k^*(i,j) = \frac{d_k(i,j)}{\sqrt{\sum_{l=0}^{3} d_l^2(i,j)}}, \quad k=0,\ldots,3 \tag{3.3-20}$$

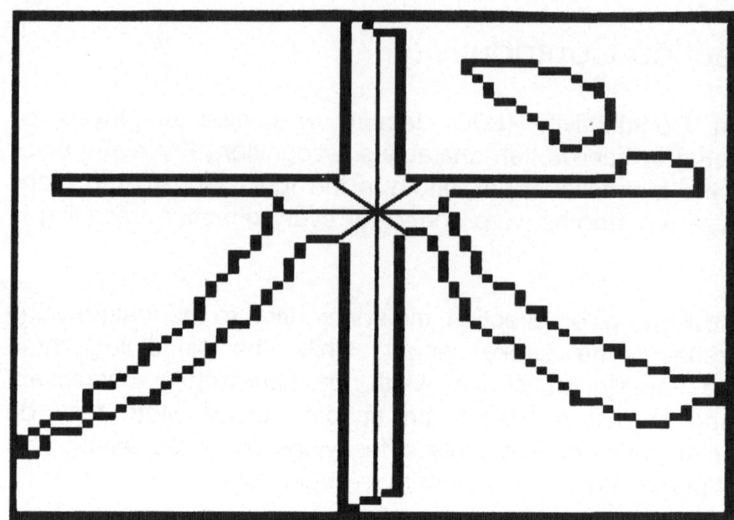

Fig. 3.3–8 Principle of the LDC feature

After that, the character pattern is divided into (n x n) regions and the average values of the normalized local features in each region are calculated so that a feature vector with 4 n^2 dimensions can be extracted.

The LDC feature was used in conjunction with 4–SDF in the first recognition stage. A 17th order classification rate of 99.99 % was reported [Hag 81].

3.35 Stroke Proportion

The Stroke Proportion feature (SP4) developed in the TECHIS project is a derivation of the LDC feature that neutralizes the influence of stroke width. When stroke width varies strongly between different instances of the same character class, only the largest local feature remains approximately constant because it varies with stroke length (instead of with stroke width, as the remaining local features do).

For this reason, LDC was modified to overcome this deficiency. The character pattern is divided into four stripes each in horizontal and vertical directions. On each stripe, four values are calculated, each of which is proportional to the number of black pixels which lie on strokes with a specific direction. Like in LDC, the bidirectional run lengths are evaluated for every black pixel in order to find the direction with maximum run length. Each pixel is then assigned the number of the maximal direction, while the run length itself is neglected:

$$f_1(n) = \sum_{i=16n+1}^{16n+16} \sum_{j=1}^{64} b(i;j) \qquad (3.3-21)$$

$$f_2(m,n) = \sum_{i=16n+1}^{16n+16} \sum_{j=1}^{64} \left\{ b(i;j) \mid \text{for each } (i,j) \; d_m(i,j) = \max_k d_k(i,j) \right\} \qquad (3.3-22)$$

$$\text{H-SP4}(m,n) = \frac{f_2(m,n)}{f_1(n)} \qquad n = 0, \ldots, 3 \qquad (3.3-23)$$

$$f_3(n) = \sum_{j=16n+1}^{16n+16} \sum_{i=1}^{64} b(i;j) \qquad (3.3-24)$$

$$f_4(m,n) = \sum_{j=16n+1}^{16n+16} \sum_{i=1}^{64} \left\{ b(i;j) \mid \text{for each } (i,j) \; d_m(i,j) = \max_k d_k(i,j) \right\} \qquad (3.3-25)$$

$$\text{V-SP4}(m,n) = \frac{f_4(m,n)}{f_3(n)} \qquad n = 0, \ldots, 3 \qquad (3.3-26)$$

There is a relation among the elements of SP4

$$\sum_{m=0}^{3} H\text{-}SP4(m,n) = 1$$

$$\sum_{m=0}^{3} V\text{-}SP4(m,n) = 1 \qquad n=0,\dots,7 \qquad (3.3\text{-}27)$$

so the dimension of the feature vector can be reduced from 32 to 24 with no loss of information.

To evaluate the stability of SP4 in comparison with LDC, the two features were tested over the same character set of N character categories. Each category is composed of M character samples with different stroke thickness. The stability is defined as the average distance among character samples of the same category divided by the average distance among different categories. The experimental results show that the stability is 18.25 % for SP4 and 35.21 % for LDC. The SP4 feature alone yielded a recognition rate of 88.85 % on CTX1 (see fig. 3.3–9 for the classification results).

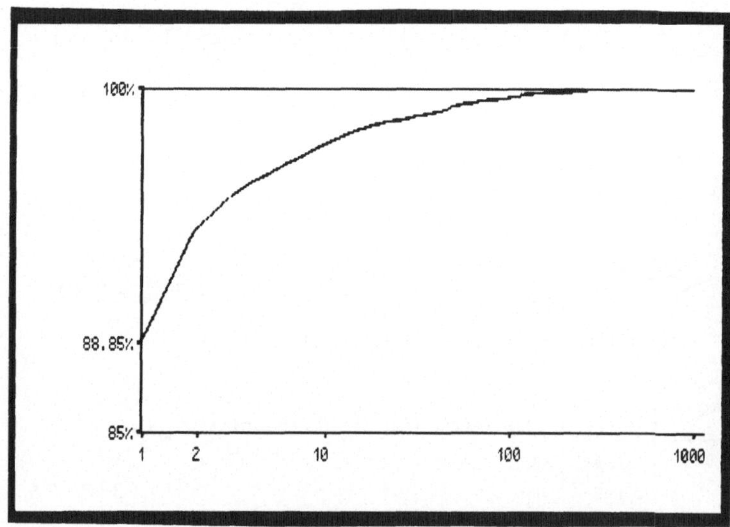

Fig. 3.3–9 SP4 classification results

3.36 Black Jump Distribution in Balanced Subvectors

The Black Jump Distribution in Balanced Subvectors (BJD–BS) developed in the TECHIS project is derived from the 4–SDF feature (see 3.33) but enhanced in two ways: balanced subfields are used instead of fixed ones, and the number of black jumps in each subvector is standardized against the total number of black jumps in the vector.

In order to localize the strokes of a character, the pattern is divided into subfields. This process can be easily demonstrated if not the matrix, but the corresponding vector is split up in the two or four main directions.

A frequent approach is to divide the (m x n) character pattern B in subfields of fixed size. This can be done by calculating the required size

$$l = \frac{m\,n}{k} \qquad (3.3\text{-}28)$$

(where k is the number of subfields) and splitting the corresponding vector of the image in k segments of length l.

Fixed subfields require the character to be approximately centered in the matrix. Noise pixels outside of the logical character boundaries, that have erroneously been included in the character segment, can influence the position of the character. Subfields that are not of fixed size but are balanced respective to blackness offer a solution to this problem.

The subfields of a character vector are defined as **balanced** if all selected subvectors contain the same number of black pixels.

By introducing balanced subfields, the effect of horizontal or vertical shifts of the character inside the binary image, as well as variations in character size, on the computed feature values is minimized.

The division of a character matrix into subfields is facilitated if the two–dimensional matrix structure is transformed into a one–dimensional vector.

Let B={b(i;j)} be a square binary image with m rows and m columns. The components (pixels) are limited to the values 0 (:white) or 1 (:black). By the following transformations, the positions of the pixels in B are arranged into a vector $\underline{v}_{\mu}$ with the components $v_{\mu}(k)$, k=1,...,m^2, in four different ways:

$$\underline{B} \in \mathbb{R}^{m \times m} \longrightarrow \underline{v}_\mu \in \mathbb{R}^N \quad \text{with } N := m^2 \tag{3.3-29}$$

in components

$$b(i;j) = v_\mu(k_\mu[i;j]) \quad \mu = 1,\dots,4 \tag{3.3-30}$$

The following four transformations are used (see also fig. 3.3–12):

(1) The arrangement of the components into the vector $v_1(\cdot)$ is built up column-wise:

$$k_1[i;j] = (j-1)m + i \quad i,j = 1,\dots,m \tag{3.3-31}$$

i.e. the components are given by

$$v_1(1) = b(1;1), \dots v_1(m) = b(m;1), \dots,$$
$$v_1(2m) = b(m;2), \dots, v_1(N) = b(m;m). \tag{3.3-32}$$

(2) The arrangement into the vector $\underline{v}_2(\cdot)$ is built up row-wise (see fig. 3.3–10):

$$k_2[i;j] = (i-1)m + j \quad i, j = 1,\dots,m \tag{3.3-33}$$

i.e. analogous with (1), but with swapped i and j;

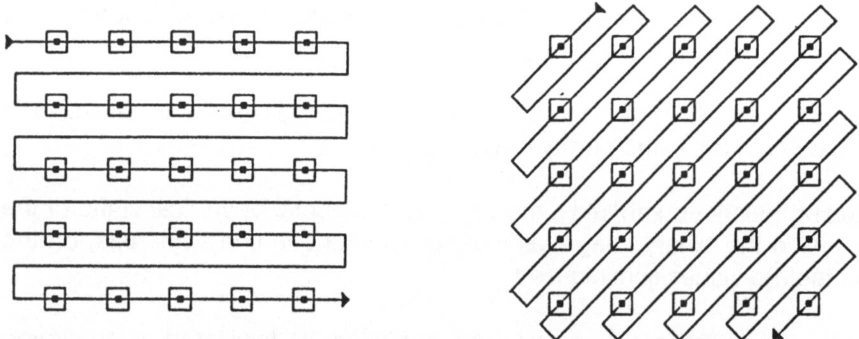

Fig. 3.3–10/11 Vectorization in horizontal and diagonal directions

(3) The vector $\underline{v}_3(\cdot)$ is built up from components at positions diagonally under 225° in the image matrix B (see fig. 3.3–11):

$$k_3[i;j] = \begin{cases} \frac{1}{2}(i+j-2)(i+j-1)+1 & \text{for all } i+j \le m+1 \\ \frac{1}{2}m(m+1) + \frac{1}{2}(3m+1-i-j)(i+j-2-m) + m - j & \text{for all } i+j > m+1 \end{cases} \tag{3.3-34}$$

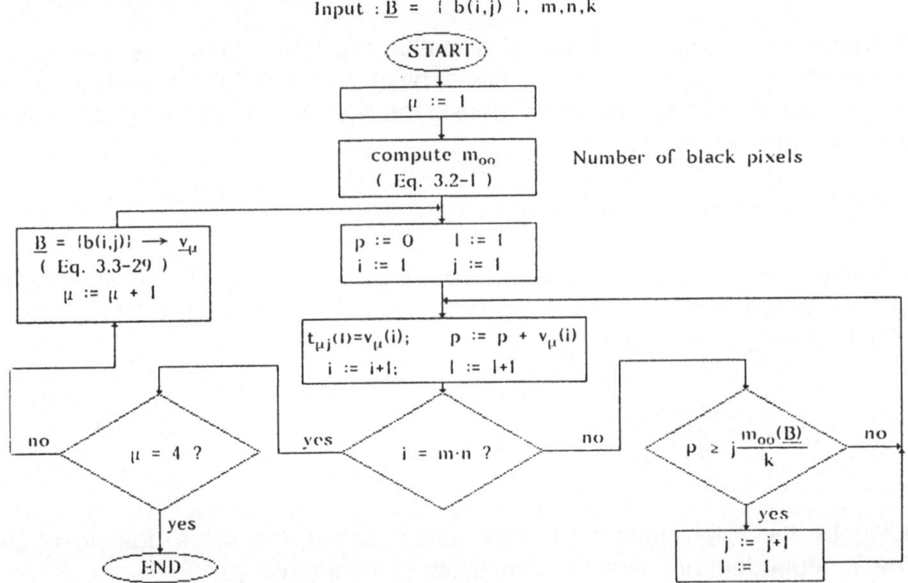

Fig. 3.3–12 Division in balanced subvectors

(4) computation of $v_4(\cdot)$ analogous to (3), where i is replaced by u:=m–i+1, i.e. diagonally under 315°:

$$k_4[i;j] = \begin{cases} \frac{1}{2}(u+j-2)(u+j-1)+u & \text{for all } u+j \le m+1 \\ \frac{1}{2}m(m+1)+\frac{1}{2}(3m+1-u-j)(u+j-2-m)+m-j & \text{for all } u > m+1 \end{cases} \quad (3.3-35)$$

Note that the vectorization can be done virtually without actually copying the pixel values. If transitions are to be counted, it is equivalent and sufficient to track the matrix pixels in the vectorization order.

The number k of subvectors (in the TECHIS project, k=8 is used) can be freely chosen under the condition (see also 3.21)

$$1 \le k \le m_{oo}(\underline{v}) \quad (3.3-36)$$

and then the number of black pixels in the balanced subvectors t_i is given by the smaller integer q of the value

$$\frac{m_{oo}(\underline{v})}{k}, \quad \text{i.e. } q = \text{int}\left[\frac{m_{oo}(\underline{v})}{k}\right] \quad (3.3-37)$$

In a typical vector representation of a matrix, the four different vectors v_μ of a binary image all have the same number of black pixels, but the dimension of the balanced subvectors $\underline{t}_{\mu i}$ can differ considerably. After partitioning, the vector equals the chained subvectors $\underline{t}_{\mu i}$:

$$\underline{v}_\mu = [\underline{t}'_{\mu 1}, \ldots, \underline{t}'_{\mu n}]' \qquad (3.3\text{-}38)$$

The feature values $x_\mu(j)$ of the subvectors $\underline{t}_{\mu j}$ ($j=1,\ldots,k$; $\mu=1,\ldots,4$) specify the proportion of the number of black jumps in the subvector to the total number of black jumps in $\underline{v}_\mu$ (see also 3.24):

$$x_\mu(j) = \frac{N_J(\underline{t}_{\mu j})}{N_J(\underline{v}_\mu)} \qquad (3.3\text{-}39)$$

Dividing by the total number of black jumps scales the value domain of the feature independent of character complexity to the interval $\{0,1\}$.

If $q\,k \leqslant m_{00}(\underline{v})$, then a rest of the original vector is not assigned to a subvector and neglected in feature computation, because this last feature correlates with the other features.

A vector $\underline{v}_\mu$ delivers k features and the total number of features for $\mu=1,\ldots,r$ is given by $N_X = k\,r$.

Additionally, it is also true that

$$\sum_{j=1}^{k} x_\mu(j) \leq 1. \qquad (3.3\text{-}40)$$

This feature proved to be relatively robust against noise, variations in size, and other typical constraints. A recognition rate of 98.3 % with CTX1 and a "re–recognition" rate of 99.5 % with GB1 were measured (for detailed results, see below 5.43).

3.37 Total Stroke Length

Because blackness is almost directly proportional to stroke width, it can be divided by the stroke width (see also 3.22)

$$TSL = m_{oo} / W \approx (m_{oo} - Q) \qquad (3.3-41)$$

to result in a feature TSL that is almost independent of stroke width and much stabler than blackness. TSL is the total length of the strokes in a character, so it is an ideal measure of the character complexity. Fig. 3.3–13 illustrates the distribution of stroke length of the 3755 Chinese characters in GB1. Because TSL returns equal values for considerable sets of tested reference characters, it may only be used in pre-classification: if the absolute difference between the stroke lengthes of input pattern and a reference pattern is less than a certain threshold, the reference is selected as a candidate for fine classification.

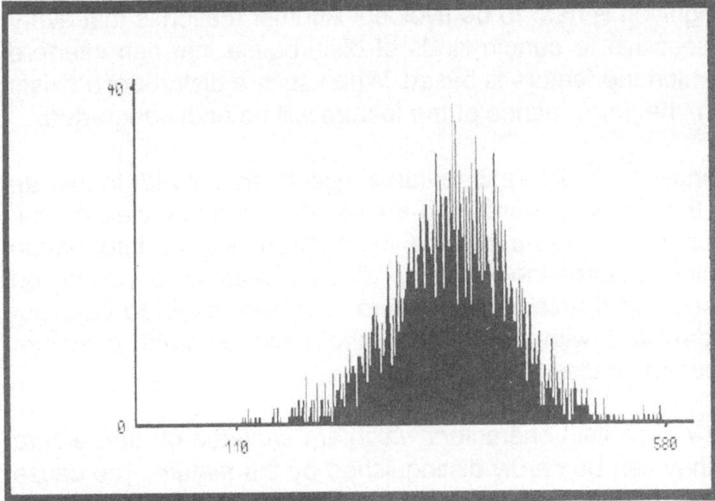

Fig. 3.3–13 Distribution of the TSL Feature

3.4 Combination of Features

3.41 Principles of Feature Combination

Although a lot of character features have been proposed and tested for Chinese character recognition, almost none of them can provide a sufficient precision when it is utilized alone, especially for documents with low print quality as are encountered in practice. This is partly due to the huge character set which results in a rather dense and crowded character distribution in the feature space.

Thus, if the feature vector extracted from the input pattern deviates from its standard vector at a distance, as is often the case under some disturbance, the input vector may become more close to a standard vector of another character category, so a misrecognition is hard to be avoided. Another reason is that every feature is particularly sensitive to certain kinds of disturbance that can interfere the measurement on which the feature is based. When such a disturbance exists in the character pattern, the performance of the feature will be badly degraded.

An appropriate combination of different features has been proved to be an effective resolution of the problem mentioned above. This idea is based on the fact that the feature combination can provide more discriminatory information and more stability against disturbance. Because different features are obtained by different measurements of the character pattern, it is reasonable to consider that they are little correlated with each other and have different character distributions on their feature spaces.

Every feature has its own "similar" characters which are crowded on the feature space so closely that they can be hardly distinguished by the feature. The cause of the problem is that the differences among these characters can not be distinctly expressed by this kind of feature. However, the "similar" characters for a certain feature may be very dissimilar for another feature, that is, the characters are likely to be far apart from each other in another feature space, therefore they may not be confused any more if the two features are employed together.

Another advantage of the feature combination lies in the improvement of the stability against disturbance. In character samples read from practical documents, there may exist various kinds of disturbance, such as blurred stroke, broken stroke, stroke width variation, character rotation, character position deviation and random noise. However, it is seldom seen that all of these disturbances occur together in a character sample. On the other hand, different features are usually sensitive to different disturbances. A set of features sensitive to different disturbances is called co–complementary, i.e. the weak points of each feature are compensated by other features. If co–complementary features are utilized jointly and the decision is made according to all of these features, a precise recognition can be expected even for disturbed character samples.

Features can be combined either in separate stages of the recognition process (see 4.13) or by joining feature values in a weighting formula. These aggregated features produce new values that can be used in preclassification. As a very simple example, the values of the peripheral features ET1 and ET2 (see 3.32) can be added to exclude candidates with heavy deviations beforehand.

As another example, the stroke density feature SDF (see 3.33) expresses the number of strokes in each region, while the features DT12 and DT23 (see 3.32) reflect the relative positions of most of these strokes. Thus it seems reasonable to utilize the three features jointly. The features can be extracted easily and require a storage space of 128 byte per character. Experiments conducted with this feature combination on the test data GB2 (see 5.41) showed a "re–recognition" rate of 99.9 % for this combination. Tests with text files resulted in recognition rates of 95.2 % for somewhat blurred characters up to 99 % for clear characters, or 97.8 % on the average.

3.42 A Successful Combination

Another recognition method based on feature combination has been developed in the TECHIS project. The employed features are ET1, ET2, SP4 (see 3.35) and 4–SDF (see 3.33). In order to investigate the performance of these features, an experiment has been carried out for the recognition of 7,589 character samples. The experiment revealed that none of these features can yield a recognition rate as high as 99 % when they are utilized alone. The more important experimental result is that the misrecognized characters, just as expected, were generally not the same for different features. This discovery confirms the conjecture that the four features are mutually independent, therefore a combination of them can give better results.

With regard to stability against disturbance, the effect of the feature combination is illustrated in fig. 3.4–1 to 3.4–6. Each figure shows a two–dimensional distribution of 7,589 character samples represented by black dots. Every character sample is located in a manner that its abscissa indicates the distance between the character sample and its reference in a feature space, and its ordinate indicates the distance in another feature space. For a noisy character sample, the more sensitive to noise a feature is, the greater distance will exist in its feature space.

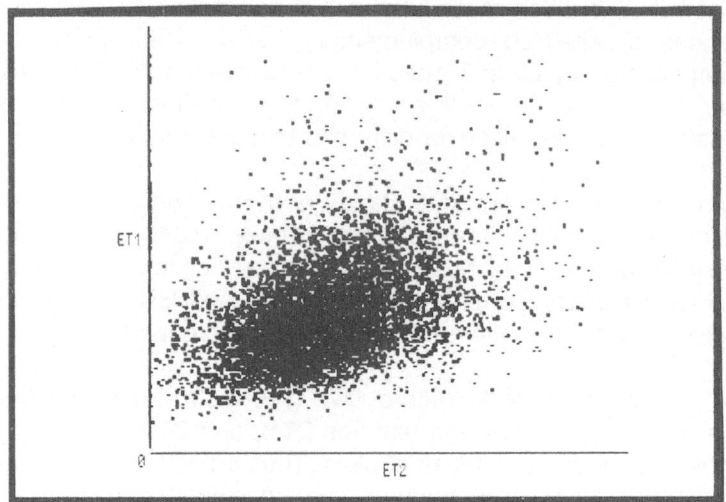

Fig. 3.4–1 Distance distribution ET1/ET2

To find out the co-complementary features, the distance distributions of all possible feature pairs among the four features are provided. In fig. 3.4–3, for example, the ordinate indicates the distance of ET1 and the abscissa indicates the distance of 4–SDF. It is well known that the feature 4–SDF is extremely unstable with respect to blurred strokes, so most black dots lying on the right area of the figure represent the character samples with blurred strokes.

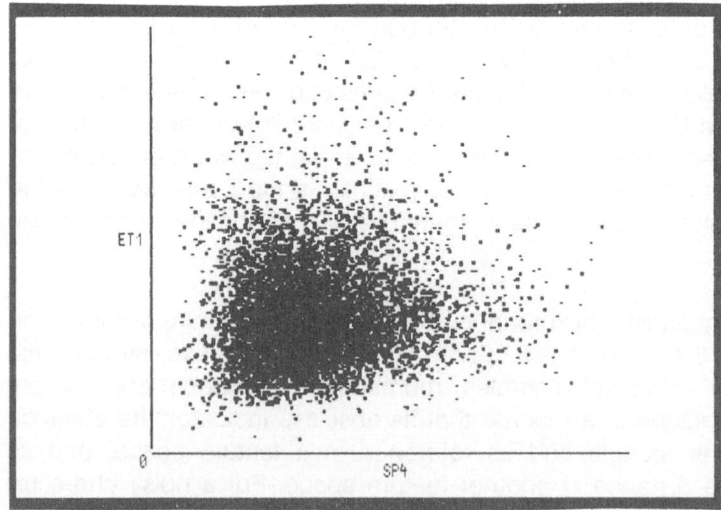

Fig. 3.4–2 Distance distribution ET1/SP4

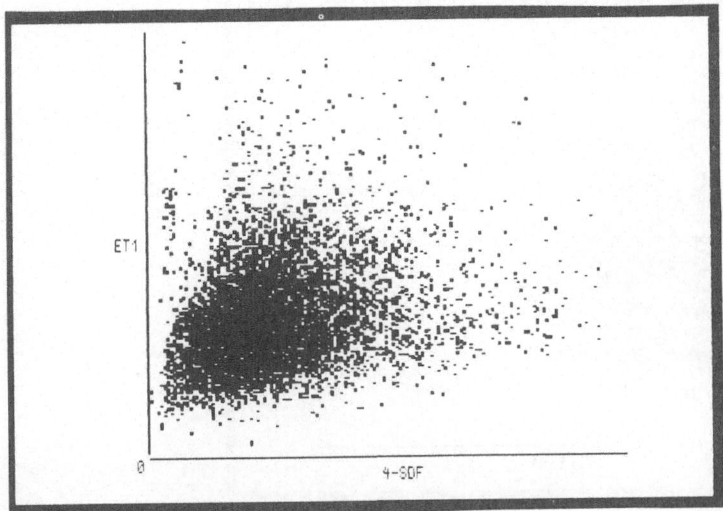

Fig. 3.4-3 Distance distribution ET1/4-SDF

Notice that most of these dots are situated on the lower area of the figure, which is understandable because the feature ET1 is fully stable against the blurred strokes. On the other hand, ET1 is more sensitive to broken strokes and position deviations than 4-SDF, that is why a number of black dots are located on the left-upper corner of the figure. The worst case appears on the right-upper corner of the figure where only few dots lie, which means that both features are

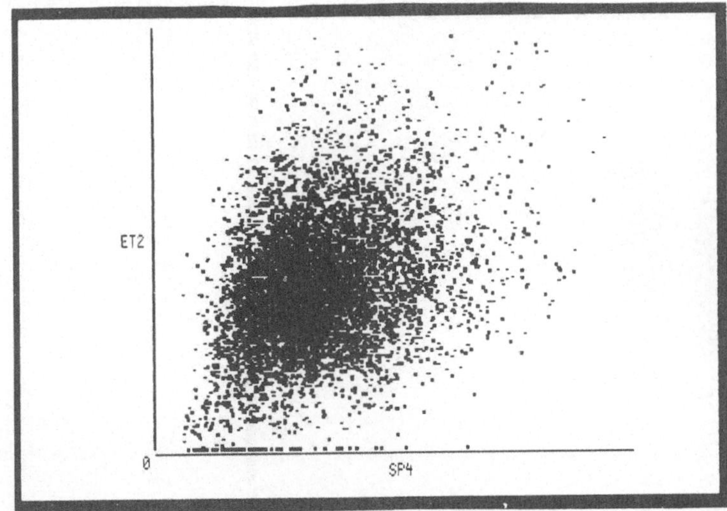

Fig. 3.4-4 Distance distribution ET2/SP4

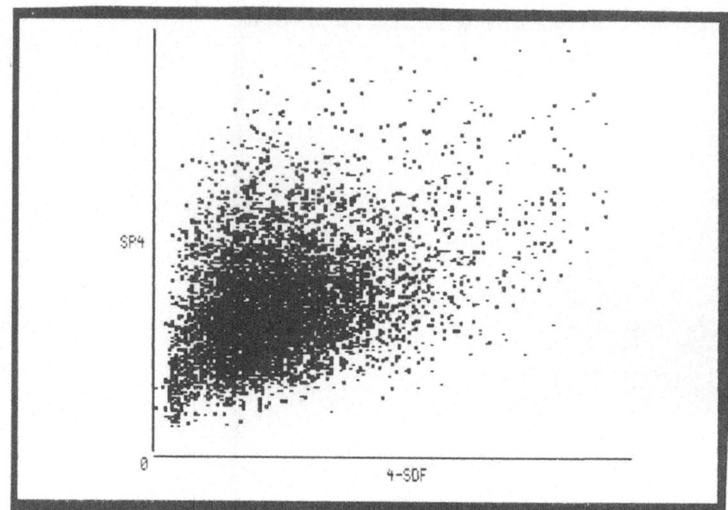

Fig. 3.4–5 Distance distribution SP4/4–SDF

degraded by serious disturbance. Even in this case, recognition of these character samples is not doomed to failure because some other feature, say SP4, may probably not be influenced by the disturbance.

The most distinctive diagram is fig. 3.4–6 which presents a striking contrast to the others. In this figure, the ordinate indicates the distance of ET2 and the abscissa indicates the distance of 4–SDF. It can be observed that all black dots

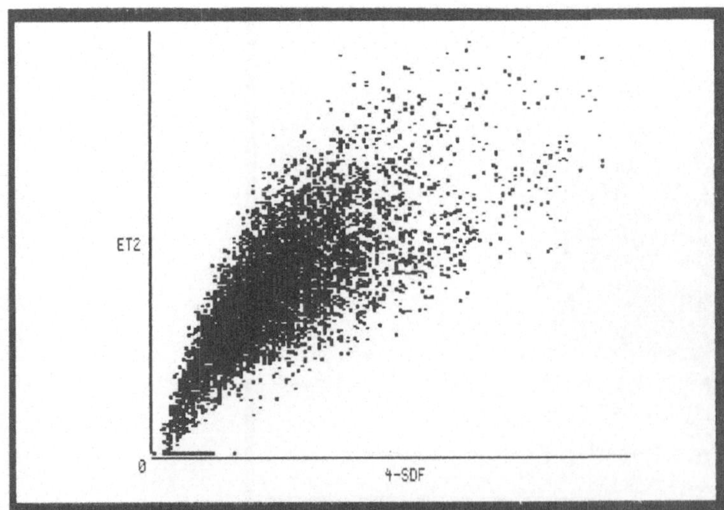

Fig. 3.4–6 Distance distribution ET2/4–SDF

are distributed in close vicinity to the diagonal, as if the distances measured in the two feature spaces were correlated.

With respect to the feature extraction methods, ET2 and 4–SDF are completely different features. ET2 relates to stroke positions, while 4–SDF depends on stroke numbers, therefore little correlation between them would be expected. However, these two features are not co–complementary, because they are sensitive to the same kind of disturbance: blurred strokes. The more serious a character pattern is blurred, the greater the distances measured in both feature spaces become. This is the cause of the strange distribution in fig. 3.4–6.

From the six distributions shown, we can draw the conclusion that the four features are co–complementary except for the fact that ET2 and 4–SDF suffer from the same disturbance, so a considerable improvement on recognition performance can be expected if the four features are appropriately combined. When a recognition method based on the feature combination is concerned, the main problem will be how to use the combined features effectively.

To reduce the required computational load, a three–stage recognition procedure was adopted in an experimental recognition system developed in the TECHIS project. In the first stage, several rough features, such as the overall stroke length, are used to select about one eighth of the total categories. In the second stage, the input sample is compared with the standard patterns of each selected category according to feature ET1 and feature ET2, respectively. The four character categories with the least distances either to ET1 or to ET2 are selected as candidates, therefore the total number of candidates can not exceed eight. The classification method developed for this feature combination will be described in section 4.3 below.

3.5 Structural Analysis

An entirely different approach to the recognition of Chinese characters was first proposed by William Stallings [Sta 72]: syntactic pattern recognition. In this category, the structural properties of Chinese characters are more taken into account. The character matrix is searched sequentially for connected groups of black pixels that are supposed to represent strokes or components. The contour of each of these black groups is traced as a closed curve.

By eliminating all contour points of insignificant directional change, the structure of the black group can be reduced to a graph that consists only of turning and touching points. The directional changes are classified into eight direction classes. Twelve direction classes have also been proposed [He 87]. This classification allows the topological description of the black group to be coded very concisely, while abstracting from size and proportions of the graph (see fig. 3.5–1). The length of the directional code varies depending on the complexity of the graph.

If a character consists of more than one connected graphs, the relation between graphs is described using a two-dimensional graphic syntax [Ran 65]. Stallings reported a recognition rate of 89 to 95 %.

From the viewpoint of character structure, the essential characteristics of a Chinese character lie in the strokes of which the character is composed and the

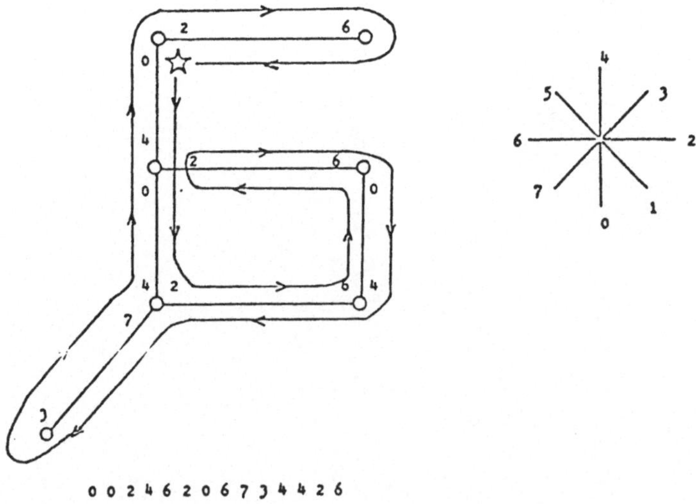

Fig. 3.5–1 Stroke direction coding

connective and positional relations among those strokes. Since the structural characteristics are less sensitive to the shape variations caused by handwriting, the stroke analysis was and still is an attractive and effective approach to handprinted Chinese character recognition.

In a recognition system based on stroke analysis, the main problem is how to decompose an image of a Chinese character into a set of fundamental strokes. Taking Zhang's work as an example [Zhang 87], the typical procedures involved in stroke extraction can be described as follows.

3.51 Thinning Algorithm

Most stroke analysis methods are based on thinning preprocessing, that is, converting the character image into a character skeleton by means of iteratively detecting and eliminating contour points (black points lying on the contours of strokes). The inherent drawback of thinning processing is that it often generates noisy and distorted skeletons, so some methods have been suggested which can extract strokes directly from the original character image. Nevertheless, the thinning method is still attractive because it can greatly simplify the stroke extraction.

As a general processing step for digital images, a number of thinning algorithms have been proposed [Tam 78]. The thinning algorithm presented here is performed by examining the black pixels one by one to find out the contour points that can be removed. A pixel $b(P_0)$ in a position P_0 and its eight neighbouring pixels are labeled in the following way:

$P_4\ P_3\ P_2$
$P_5\ P_0\ P_1$
$P_6\ P_7\ P_8$

During the scanning of the character image, the pixel $b(P_0)$ is marked as a removable contour point if the following three conditions are fulfilled:

$$(1)\quad [b(P_0)=1] \wedge \sum_{i \in S} \Big[|b(P_0) - b(P_i)| \Big] \geq 1 \qquad (3.5-1)$$

$$(2)\quad \sum_{k \in S} \Big[\overline{b(P_k)} - \overline{b(P_k)}\ \overline{b(P_{k+1})}\ \overline{b(P_{k+2})} \Big] = 1 \qquad (3.5-2)$$

$$(3) \quad \sum_{i=1}^{8} b(P_i) > 1 \tag{3.5-3}$$

$$\text{where } P_9 = P_1 \text{ and } \overline{b(p_k)} = 1 - b(p_k). \tag{3.5-4}$$

$$s = \{1, 3, 5, 7\}$$

The first condition detects the contour point, the second condition keeps the connectivity of the image, while the third condition prevents end points of thinned lines from being removed. A pixel which has been marked as a removable contour point is regarded as a black point in calculation of the first condition and as a white point in calculation of the second and the third conditions.

After all pixels have been examined, the marked points are changed to white points, so that a layer of contour points is removed and the strokes become thinner. If there is no marked point, it indicates that the image has been thinned into a skeleton, otherwise the thinning operation is repeated to remove the next layer of contour points.

3.52 Feature Point Detection

On a character skeleton, the topological property of each black point can be measured according to its eight neighbouring points. The crossing number of a black point is here defined as

$$N(P_0) = \frac{1}{2}\left[\sum_{k=1}^{8} | b(P_k) - b(P_{k+1}) | \right] \tag{3.5-5}$$

The possible values of crossing number are from 0 to 4, which have the following meanings:

$$N(P_0) = \begin{cases} 0 & P_0 \text{ is an isolated noise point;} \\ 1 & P_0 \text{ is an end point;} \\ 2 & P_0 \text{ is a connective point;} \\ 3 & P_0 \text{ is a branch point;} \\ 4 & P_0 \text{ is a cross point.} \end{cases}$$

In Chinese characters, most strokes are straight lines in different directions. The others can be approximately regarded as folding lines. Because a section of straight line is determined by its end points, it may be concluded that the structural information of a Chinese character concentrates on some so–called feature points. There are four kinds of feature points: end point, branch point, cross point, as well as corner point. The feature points of the first three kinds can be determined directly by crossing number, while the corner points are a subset of the connective points, which will be detected during the procedure of stroke segmentation.

3.53 Stroke Tracking and Stroke Segmentation

From a feature point, it is very easy to track a section of a stroke along the skeleton until another feature point is encountered. A stroke section is represented as a sequence of points

$$L = (P_1, P_2, \ldots, P_n) \qquad (3.5-6)$$

where P_1 and P_n are feature points, the others are connective points. By this way, all strokes starting and terminating at feature points can be found out from the character skeleton.

In general, it is possible to segment a stroke into several approximate straight lines. The segmentation is performed by detecting corner points where the curvature of the stroke changes abruptly. The variation of the curvature at a point P_i can be indirectly measured by the local angle $\Psi(P_i)$. Fig. 3.5–2 illustrates the calculation method of $\Psi(P_i)$, where r is a constant.

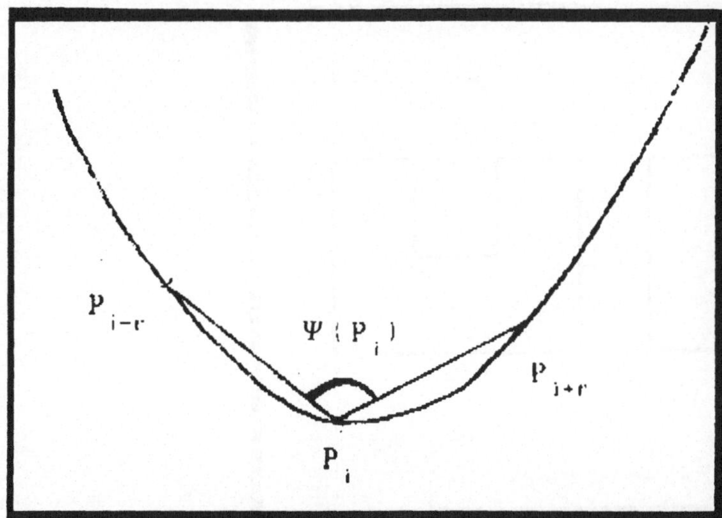

Fig. 3.5–2 Calculation of the local angle

The segmentation algorithm can be described as a recursive procedure:

(1) Let $L=(P_1,P_2,...,P_n)$ be the stroke to be segmented. If $n < 2r$ then L is approximated by a straight line; go to EXIT. Otherwise:

(2) Find out the maximal local angle on the stroke

$$\Psi(P_k) = \max\{ \Psi(P_i) \mid r < i < n-r \} \quad (3.5-7)$$

(3) If $\Psi(P_k) > \theta$, then L is approximated by a straight line, go to EXIT. Otherwise:

(4) P_k is a corner point, L is segmented into two sections of stroke

$L1=(P_1,P_2,...,P_k)$

$L2=(P_{k+1},P_{k+2},...,P_n)$

(5) Perform the segmentation of L1;

(6) Perform the segmentation of L2;

(7) EXIT.

where r and θ are constants.

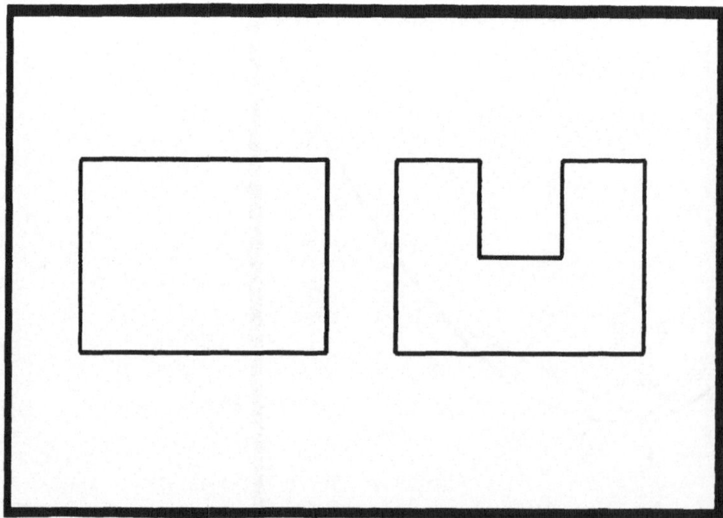

Fig. 3.5-3 Examples of closed strokes

Sometimes there may be perfect closed lines in the character skeleton (see fig. 3.5–3 for some examples). The closed lines cannot be found out during the procedures mentioned above, because they are composed only of connective points. This kind of strokes is extracted by scanning the character skeleton. When encountering a point which has not been tracked, we can track from this point until return to the same point. The strokes found in this way are also broken into several straight lines by the segmentation procedure.

3.54 Combination of Segments

Through the procedures of stroke tracking and segmentation, the character skeleton is decomposed into a number of straight line sections as shown in fig. 3.5–4. These sections are basic elements of strokes, some of which will be put together to synthetize various strokes. The first step of the stroke synthesis is the combination of straight lines, that is, to combine the sections into a set of straight lines which should be stretched as long as possible. For example, the L1 and L2 in fig. 3.5–4 should be linked together at the branch point. L4 and L5, as well as L3 and L6, should be linked together at the cross point. L7 is not connected with L6.

For an ideal character skeleton, it is not difficult to find out the segments to be combined. At every branch point and cross point, we can test all possible pairs of segments to determine whether they are in good continuity or not. The continuity of a pair of segments is measured by the included angle between them. If the angle is greater than a certain threshold, the two segments can be linked together.

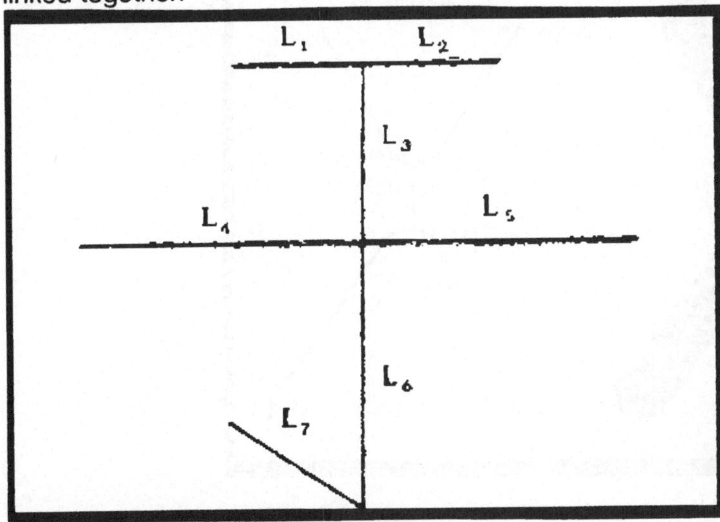

Fig. 3.5–4 Skeleton decomposition

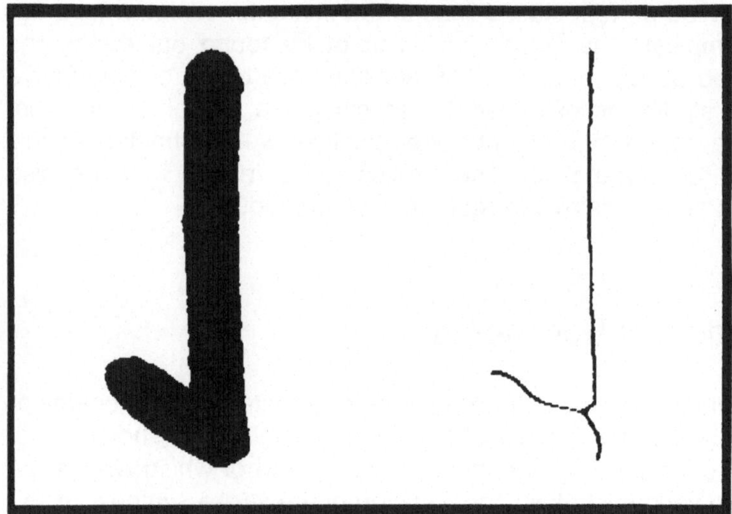

Fig. 3.5–5 Distorted corner

The largest difficulty involved in the segments combination is the discrimination of pseudo sections caused by the thinning processing. Although thinning processing can simplify the drawing line tracking, it generates usually distorted character skeletons, especially when the strokes are relatively thick. The typical distortions are distorted corners as shown in fig. 3.5–5 and the distorted "X" type intersections as shown in fig. 3.5–6.

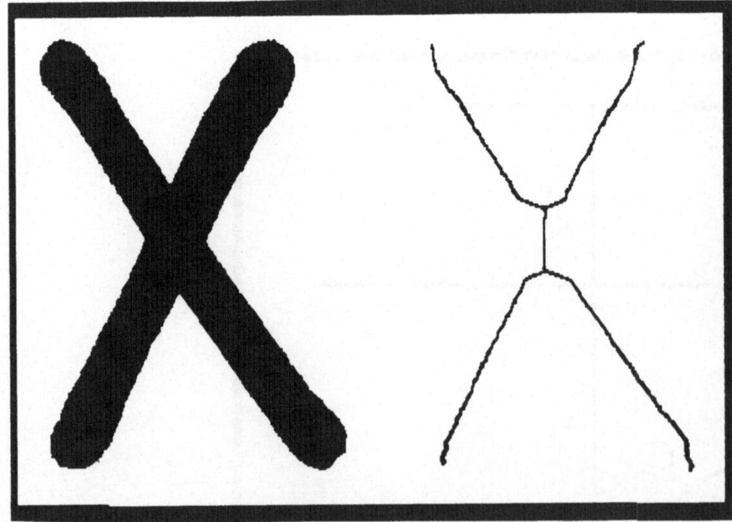

Fig. 3.5–6 Distorted intersection

From fig. 3.5–5, it can be seen that a noise segment like a "whisker" is generated in the opposite direction of the corner. The length of a whisker depends on the thickness of the stroke and is usually shorter than the normal segments of strokes. The method of removing whiskers is based on the characteristics of this kind of distortion. For every branch point, the three segments connected to the branch point are compared according to their lengthes. The longest one is labeled L1 and the others are labeled L2 and L3. Under the following three conditions, L1 is detected as a pseudo segment:

(1) One of the terminals of L1 is an end point;

(2) The length of L1 is less than a threshold;

(3) The included angle between L2 and L3 is less than a threshold.

Once a whisker is detected, it is eliminated and the branch point is changed into a corner point. When two strokes intersect with an acute angle, the thinning process will generate two branch points on the intersection area instead of a cross point. Between the two branch points there is a pseudo segment which looks like a "neck". A typical distorted intersection is shown in fig. 3.5–6. This kind of distortion is removed by testing every segment according to the following three conditions:

(1) Both terminals of the segment are branch points;

(2) The length of the segment is less than a threshold;

(3) Among the other segments connected to the two branch points, there are two pairs of segments in good continuity.

3.55 Stroke Synthesis

After the segments combination, a set of straight lines has been obtained. Some of these straight lines are already strokes, while others are stroke elements which will be further combined into complex strokes. In Chinese philology, a stroke is defined as the trace generated by one movement of the pen. However, in the research field of Chinese character recognition, almost every distinct recognition method has its distinctive definition of strokes. From the viewpoint of structural analysis, the strokes are the primitive elements composing the characters. Therefore the strokes are often defined as some simple subpatterns that are easy to be extracted and stable against noise and distortions.

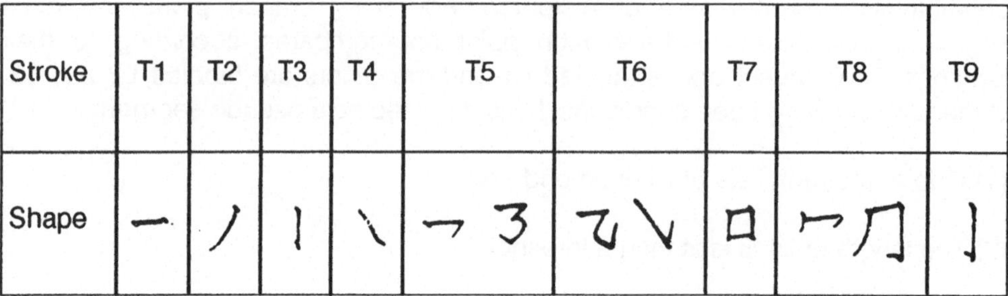

Stroke	T1	T2	T3	T4	T5	T6	T7	T8	T9
Shape									

Fig. 3.5–7 Fundamental strokes

In the recognition method described here, nine types of subpatterns are defined as fundamental strokes. The typical shapes of these stroke types (T1...T9) are illustrated in fig. 3.5–7.

These nine types of strokes are the most frequently used subpatterns in Chinese characters. The first four types are straight lines quantized into four directions. T5 and T6 summarize most of the curvic strokes in Chinese characters. A curvic stroke is composed of several segments linked by corner points.

There are two terminal segments in a curvic stroke. If the terminal segment with the lower position turns left, the curvic stroke has the stroke type of T5, otherwise it belongs to T6. T7 and T8 are two simple character components that are synthesized by two or three simpler strokes. T9 is a special type of curvic stroke, which is often confused with T3 in handwriting.

3.56 Structural Description

A Chinese character is characterized by the strokes of which it is composed, and by the structural relations among the strokes. As compared with alphanumerics, Chinese characters are much richer in structural information. For instance, the stroke type, stroke length and stroke position can be, and sometimes have to be, used to represent properties of a stroke. As to the relations among strokes, there are positional relation, cross relation, connect relation, enclose relation, and so on. For the task of automatic recognition, it is too redundant if the complete structural information is evaluated. So it is possible to select some stabler relations to describe the structure of the characters.

Among the various structural relations, the cross relation is most stable against the shape variation caused by handwriting. According to the cross relation, a set of strokes are defined as a crossing group if the strokes are intersected together.

In many Chinese characters, such as 围 and 胃 , there are some strokes enclosed by a loop(T7). The enclose relation can be reliablly extracted from the character skeleton, therefore it is selected as a structural relation used in character description.

The characteristic of a Chinese character is represented by a descriptive code. A descriptive code is composed of three sections, which are called crossing section, enclosing section and independence section. These sections will now be described.

The general format of the **crossing section** is shown as follows:

$$K = \left[\Phi_1 \; \Phi_2 , \ldots , \Phi_k \right] \qquad (3.5-8)$$

where K is the number of crossing groups in the character. When K = 0, the items between the parenthese do not exist in the code. The i–th crossing group is described by Φ_i which has the form of

$$a_i \; b_i \; s_1 \; s_2 , \ldots , s_{a_i} \qquad s_k \in \{T_i\} \qquad (3.5-9)$$

where a_i is the number of strokes in the group, b_i is the number of intersections, while $s_1..s_n$ represent the types of those strokes. The strokes may be arranged in arbitrary order, because only the proportion of each type of stroke is considered when calculating the similarity between two descriptive codes. The order of Φ_1 ... Φ_k is arranged according to the stroke number, intersection number, as well as the stroke proportion of the crossing group.

Taking the character 菌 as an example, the crossing section of its descriptive code is "2321332113". This string of digits is equivalent to the following description: "There are two crossing groups in the character(2). The first crossing group includes three strokes(3) and two intersections(2), the three strokes are T1,T3 and T3(133), respectively. There are two strokes(2) and an intersection(1) in the second crossing group, the type of the two strokes are T1 and T3(13)."

The **enclosing section** has the following format

$$d_1\left[\, d_2 d_3 (s_1 s_2, \ldots, s_{d_3}) \,\right], \quad s_k \in \{T_i\} \qquad (3.5-10)$$

In this section, d_1 is the number of loops enclosing some strokes. If there is no such loop, the section is simply composed of a zero. d_2 is the ordinal number of the crossing group enclosed by a loop. When there is no enclosed crossing group, d_2 is set to zero. d_3 is the number of enclosed strokes other than the enclosed crossing group. When d_3 is greater than zero, the stroke types should be listed in the form of $s_1 \ldots s_n$.

The enclosing section of descriptive code of the character 菌 is "123224" which has the following meaning: "In the character there is a loop (1) enclosing some strokes. The second (2) crossing group is enclosed by the loop. Besides this crossing group, there are still three (3) enclosed strokes whose types are T2, T2 and T4(224)."

The **independence** section is used to describe the strokes that are neither included in a crossing group nor enclosed by a loop. This section takes the form of

$$n\ s_1 s_2, \ldots, s_n \qquad (3.5-11)$$

where n is the number of independent strokes. The independent section of the character 菌 is "17", which means that in this character there is only one independent stroke and the type of the stroke is T7.

character	descriptive code
一	0 0 11
我	15419262 0 224
田	12113 110 17
兴	0 0 6442124
汉	12154 0 3244
帽	12138 203111 277

Fig. 3.5-8 Examples of descriptive code

A complete descriptive code is composed of the three sections mentioned above. For example, the character 菌 has a descriptive code of "232133211312322417". Examples for descriptive codes are shown in fig. 3.5–8.

An experiment has been carried out for a character set of 500 frequently used character categories. The data used in the experiment included 700 learning samples and 1,300 test samples which were carefully written by four persons. After a descriptive code has been extracted from the input character pattern, it is compared with every reference code which represents a character category. The character category whose reference code is most similar to the input code is assigned to the input character pattern. In the experiment, a recognition rate of 95% was achieved for the test samples (false rejections 4%, substitutions 1%).

A problem of structural recognition methods is that they require character patterns of rather high quality in which no strokes should be broken and no components should touch each other irregularly. Complex characters in body text sizes, produced under suboptimal but typical printing conditions, cannot always meet this demand (see fig. 3.5–9 for a misleading skeleton). With higher-resolution input devices and very sensible preprocessing, structural recognition seems however to be feasible.

Fig. 3.5–9 Misleading skeleton

4 Classification

4.1 Principles of Classification

In the previous chapter, some feature extraction methods appropriate for Chinese characters have been treated. In the third stage of the character recognition system, feature classification methods have to be found. These methods are to divide the feature space in (pre–) classes in order to identify a single character or a group of characters.

The classifiers are required to separate the feature space into K mutually exclusive regions. This aim is pursued by classifier design, training or learning.

The partition is usually performed through discriminant functions $d_k(x)$ with the property

$$d_k(\underline{x}) \leq d_i(\underline{x}) \quad \text{for all } i \neq k \tag{4.1-1}$$

then the feature vector x belongs to the class C_k.

The discriminant functions, frequently referred to as distance functions, are determined as proper distance in a metric or norm space. In general, the distance $d(x_k,x_i)$ between two feature vectors x_k and x_i in the metric space has to satisfy the following conditions:

$$1.\ d(\underline{x}_k, \underline{x}_i) = 0 \text{ implies } \underline{x}_k = \underline{x}_i \tag{4.1-2}$$

$$2.\ d(\underline{x}_k, \underline{x}_i) = d(\underline{x}_i, \underline{x}_k) \tag{4.1-3}$$

$$3.\ d(\underline{x}_k, \underline{x}_i) \leq d(\underline{x}_k, \underline{x}_j) + d(\underline{x}_j, \underline{x}_i) \tag{4.1-4}$$

The rules are similar for a map (or norm function) from the linear vector space $\mathbb{R}^N(\mathbb{C}^N)$ in the real or complex numbers,

i.e. $||\cdot|| : \mathbb{R}^N \longrightarrow \mathbb{R}(\mathbb{C})$:

$$1.\ ||\,\underline{x}\,|| \geq 0 \text{ and } ||\,\underline{x}\,|| = 0 \text{ implies } \underline{x} = 0 \tag{4.1-5}$$

$$2.\ ||\,\alpha\underline{x}\,|| = |\alpha|\ ||x||; \qquad \alpha \in \mathbb{R}(\mathbb{C}) \tag{4.1-6}$$

$$3.\ ||\,\underline{x} + \underline{y}\,|| \leq ||\underline{x}|| + ||\underline{y}|| \quad \underline{x}, \underline{y} \in \mathbb{R}^N(\mathbb{C}) \tag{4.1-7}$$

It is clear that in the linear norm space, the connection between the two maps is given by

$$d(\underline{x}_k, \underline{x}_i) = || \underline{x}_k - \underline{x}_i || \qquad (4.1-8)$$

For classification problems, mostly the l_p norm (where $1 \leqslant p < \infty$) or the l_∞ norm are chosen. They are defined as

$$||\underline{x}||_p := \left[\sum_{i=1}^{N} |x(i)|^p \right]^{\frac{1}{p}} \qquad (4.1-9)$$

$$||\underline{x}||_\infty := \max_i |x(i)| \qquad (4.1-10)$$

Each pattern classification procedure will determine only with a certain probability to which category or class a given sample belongs. In character recognition for commercial use, a recognition rate of over 99 % is required.

Statistical pattern classification is based on the assumption that for each class the conditional density function $p_i(\underline{x}|C_i)$ and the a priori probability P_i of C_i are known. The resulting error rates in the classification problem can then be calculated.

Unfortunately, both the conditional density function and P_i are unknown in most cases. Statistical parameters have to be used: conditional expectation vectors (mean vector of a class) and conditional covariance matrices can also approximately solve the classification problem. These statistical parameters have to be estimated from the training set.

There are two approaches to classification of the feature space. The first uses reference characters as training set. Those feature vectors of reference characters form a cluster (or pre–class) in which the distance to the feature vector of the input character is smaller than a given threshold.

The second method requires certain learning techniques. To each reference character, a cluster of points in the feature space is assigned. For pre–classes, the intersections of clusters need not be empty. An input character is (pre–) classified if its feature vector belongs to one of the clusters (classes).

In character recognition, two different ways have been developed: the single character and the text (word) recognition approach. This monograph deals only with the single–character approach, while word recognition is still under investigation.

4.2 Classification Tools

4.21 Special Discriminant Function

To judge the proximity between two feature vectors, a number of distance measures have been proposed. The simplest classificator determines the distance between two feature vectors with the l_1 norm, also known as linear or city-block distance [Kit 86:68, Nie 83:211].

Let x be the feature vector of an unknown sample and x_i be the average feature vector of class i. All feature vectors consist of N elements. The l_1 norm to serve as discriminant function between x and x_i is then computed as

$$d_i(\underline{x}) = ||\underline{x}_i - \underline{x}|| = \sum_{j=1}^{N} |x_i(j) - x(j)| \quad (4.2-1)$$

The advantage of this linear distance measure lies in the fact that it involves only adding and subtracting and is thus highly speed-efficient in computers.

In practical applications, the Euclidian distance (l_2 norm) or other l_p norm measures are less helpful because square (or p) roots require considerable computation time. In the TECHIS project, linear distance was mostly employed.

4.22 Similarity

The similarity method is one of the most frequently used classification methods, see for example [Ume 79:761]. Considering that the input character pattern is represented by a n-dimensional feature vector in a form of

$$\underline{x} = [x(1), \ldots, x(N)]^T \quad (4.2-2)$$

The standard characters in the candidate set are expressed in

the same form as

$$\underline{y}_k = [y_k(1), \ldots, y_k(N)]^T \quad k = 1, 2, \ldots, m \quad (4.2-3)$$

where m is the number of characters in the candidate set.

The similarity between an input character and the k-th standard

character is defined as follows:

$$S(\underline{x};\underline{y}_k) = \frac{<\underline{x},\underline{y}_k>}{\sqrt{<\underline{x},\underline{x}> \; <\underline{y}_k,\underline{y}_k>}} \qquad (4.2-4)$$

$$\text{where} \quad <x,y> = \sum_{i=1}^{N} x(i)\,y(i) \qquad (4.2-5)$$

This similarity measure is sometimes called the correlation coefficient, if <x,y> is the (estimated) covariance of the random variables $\underline{x}$ and $\underline{y}$.

In the deterministic case, the similarity $S(\underline{x};\underline{y})$ is actually the cosine of the angle between the vectors x and y. It has two properties:

$$(1) \quad 0 \leq S(\underline{x};\underline{y}) \leq 1 \qquad (4.2-6)$$

(S is not negative because all x(i), y(i) $\geqslant$ 0).

$$(2) \quad S(\underline{x},\underline{y}) = 1, \qquad (4.2-7)$$

if and only if x = αy, where α is a positive scalar.

The "Maximum Similarity Criterion" is often used to make the decision, that is, the character category yielding the largest similarity is assigned to the unknown input character pattern. Our experiments showed, on the other hand, that the linear classificator performs a faster classification with comparable correctness, because less calculations were required.

4.23 Classification Behavior

In judging the "behavior" of a classification procedure, several cases can be distinguished. The unknown sample may be valid, that means it is represented by a result class of the recognition system, or invalid for out-of-set characters (for example, a Hebrew letter for a system that recognizes only characters from the Latin alphabet).

The recognition system has at least three kinds of reaction to a sample: returning the correct result (this is, of course, only possible for valid characters), returning an incorrect code (this is often called **substitution**), and rejecting the sample outright.

Validity and recognition reaction can be combined in five possible ways that are arranged here according to decreasing quality:

(1) Valid sample, correct code – perfect. That's how it should always be.

(2) Invalid sample, reject – also perfectly correct.

(3) Valid sample, reject – a nuisance. Should not occur too often, since it requires post-processing.

(4) Invalid sample, incorrect code – dangerous, because the substitution may pass unnoticed.

(5) Valid sample, incorrect code – extremely bad. This is the worst case in recognition and should be avoided at any cost.

By modifying the reject conditions, most cases of (5) can be changed to (3), cases of (4) would then also change to (2). If cases (2) and (4) occur often, the working character set should be revised.

4.3 Combining Distance Measures

In the last stage of recognition, the input character sample must be assigned to a character category within the selected candidate set (pre–class). This task is very difficult because mutually similar characters are usually gathered in the set. In order to provide sufficient discriminatory information, all of the four features must be utilized together. However, traditional statistical decision methods are based on the division of a single feature space instead of several feature spaces, so they cannot be directly used for the decision by different features.

Umeda [Ume 84] has reported a recognition scheme using the combination of mesh feature and peripheral feature. He used the weighted sum of the distances for the two features as the overall measurement of similarity, where the weighting value is determined experimentally. But this method is not applicable to the combination of more than two features, because it is almost impossible to determine several optimal weighting values by recognition experiments.

To cope with this problem, two kinds of similarity measurements, namely relative distance and absolute distance, have been proposed in the TECHIS project. The former is used to select a character category which is most similar to the input sample according to all of the four features, while the latter is used to determine whether the input sample is acceptable or should be rejected. At first, the distances between the input sample and each character category in the candidate set are calculated for each feature. The distance measure may be arbitrary, but the linear distance is adopted in the TECHIS system because it requires much less computation than other distance measures, such as the euclidian distance. On this basis, the relative distance of the k–th character category in the candidate set is defined as

$$\tilde{D}_k = \sum_{j=1}^{4} \frac{D_k^j - \min\{D_i^j \mid i=1,\ldots,N\}}{\max\{D_i^j \mid i=1,\ldots,N\} - \min\{D_i^j \mid i=1,\ldots,N\}}\,,$$

$$k=1,\ldots,N \qquad (4.3-1)$$

where N is the number of candidates and D_i^j is the distance of the i–th candidate for the j–th feature. Here the main concern is not to measure how similar the candidate is to the unknown sample actually, but instead to measure how similar the candidate is to the unknown sample as compared with other candidates, that is why it is called relative distance.

The actual similarity between the unknown sample and the k–th candidate is measured by the absolute distance which is defined as follows

$$\hat{D}_k = \sum_{j=1}^{4} \frac{|x(j) - x_k(j)|}{E\{|x(j) - x^*(j)|\}} \hat{=} \sum_{j=1}^{4} \frac{D_k^j}{\overline{D^{*j}}} \quad . \quad k = 1, \ldots, N \tag{4.3-2}$$

where $\overline{D^{*j}}$ is the expected value of the distance between a character sample and its reference for the j–th feature, which is estimated by the learning samples beforehand and acts as a constant in the recognition procedure. The reason of this definition lies in that the different features are obtained by different measurements with different yardsticks, therefore it is utterly meaningless to simply add up the distances for the four features.

The final decision is made according to the following rule: If the absolute distance of the candidate with the minimal relative distance is less than a threshold, the character category of the candidate is assigned to the unknown sample, otherwise the unknown sample is rejected by the system.

To evaluate the effectiveness of the feature combination, a recognition experiment has been carried out on the data of more than 20,000 characters taken from 15 pages of ordinary Chinese magazines (CTX1, see 5.41). The method has shown an exellent recognition performance even for noisy printed characters. The recognition rate was 99.92 percent for the test data.

Umeda [Ume 82:795] reported that a combination of the mesh and the peripheral feature recognized 99.45 % and classified 99.99 % up to the third order of 2,024 categories in different fonts.

4.4 Hierarchical Classification

When dealing with a larger set of classes, like Chinese characters, sequential application of one classification algorithm will most often be too time–consuming. Therefore, a hierarchy of a preclassification and a fine classification stage is commonly applied. The principle of preclassification (also called **candidate selection**) consists of defining a set of pre–classes, each pre–class being a subset of the set of object classes. For every character to be recognized, its pre–class is first determined with simple, "rough" but fast features. In the second phase of **fine classification**, only the classes in the pre–class (candidates) are considered and tested with another set of features.

The number of elements in a pre–class is supposed to be much less than the total number of classes. For Chinese characters, the whole set of classes comprises several thousand characters, while a pre–class should contain no more than 100 candidates. This leads to higher processing speed in fine classification.

Requirements to the fine classification procedure may also be eased under the condition that the features used by pre– and fine classification are not correlated. If pre–class sizes are small, more complex feature extraction methods that require much computation time can still be tolerated.

A division in more than two classification stages is also possible and can further increase recognition speed, if adequate features are chosen for the respective stages.

4.5 The Overlap Problem

At the border of a pre–class, uncertain recognition (resulting in higher error rates) poses a problem for most pre–classification procedures. In a n–dimensional feature space, a sample feature vector could in the worst case have equal distance to n pre–classes that all overlap.

In order to receive better pre–classification results, an overlap of the pre–classes is necessary: unknown characters may be assigned more than one pre–class. For precise fixing of the overlap regions, the conditional distribution densities of the pre–class or other suitable statistical parameters have to be determined.

A procedure for determining the pre–class limits and the overlap regions in the feature space may be defined as follows:

Let $A = \{ i | i \leqslant K \}$ be the index set of the character set to be recognized. If the conditional expectation values for the feature vectors of given character classes are known, the index set A may be divided into subsets A_j with K_j elements generated by a cluster analysis algorithm (for example, the minimum distance cluster algorithm). These subsets will be referred to later as pre–classes.

After the elements of the subset A_j have been determined, the conditional distribution densities $p(x|A_j)$ of the pre–classes A_j are calculated. These can be estimated from the distribution densities $p(x|i)$ of the separate characters of a pre–class:

$$p(\underline{x} \mid A_j) = \frac{1}{K_j} \sum_{i \in A_j} p(\underline{x} \mid i) \quad (\underline{x}\text{: input feature vector}) \tag{4.5-1}$$

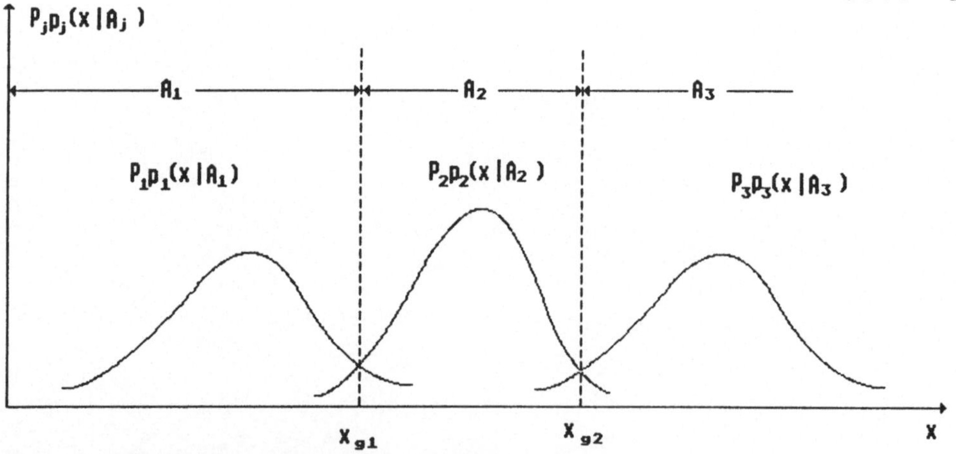

Fig. 4.5–1 Examples of overlapping classes

If the conditional distribution of the feature vectors of the pre–class A_j is a normal distribution, it can be described as

$$p(\underline{x} \mid A_j) = \frac{1}{\sqrt{|2\pi \det \underline{C}_j|}} \exp\{-\tfrac{1}{2}(\underline{x}-\underline{\mu}_j)^T \underline{C}_j^{-1} (\underline{x}-\underline{\mu}_j)\}. \quad (4.5-2)$$

where $\underline{\mu}_j$ is the conditional average vector of the pre–class A_j

$$\underline{\mu}_j = E\{\underline{x} \mid A_j\} \quad (4.5-3)$$

and $\underline{C}_j$ is the conditional covariance matrix

$$\underline{C}_j = E\{(\underline{x}-\underline{\mu}_j)(\underline{x}-\underline{\mu}_j)^T \mid A_j\} \quad (4.5-4)$$

The conditional average vector and the co–variance matrix are then estimated from a number of samples as will be demonstrated for the limits of three pre–classes and a one–dimensional feature vector:

Let P_j be the a priori probabilities of the pre–classes. Fig. 4.5–1 shows the distribution densities $P_j p_j(x|A_j)$. The pre–class limits x_{g1} and x_{g2} result from [Nie 83:159–175]:

$$P_1 p_1(x_{g1} \mid A_1) = P_2 p_2(x_{g1} \mid A_2) \quad (4.5-5)$$

$$P_2 p_2(x_{g2} \mid A_2) = P_3 p_3(x_{g2} \mid A_3) \quad (4.5-6)$$

If the a priori probabilities of all pre–classes are equal and the conditional distributions of the pre–classes are distributed normally, then the pre–class limits x_{g1}, x_{g2} can be computed from

$$x_{g1} = \frac{\sigma_2 \mu_1 + \sigma_1 \mu_2}{2(\sigma_1 + \sigma_2)} \quad (4.5-7)$$

$$x_{g2} = \frac{\sigma_3 \mu_2 + \sigma_2 \mu_3}{2(\sigma_2 + \sigma_3)} \quad (4.5-8)$$

where σ_j is the standard deviation of the features in the j–th pre–class.

The probability for a pattern to yield an incorrect pre–classification (error probability [Nie 1983, p. 174]), is

$$P_f = 1 - \sum_{j=1}^{3} P_j \int_{B_j} p_j(\underline{x} \mid A_j)\, d\underline{x} \quad (4.5-9)$$

$$B_j = \{\underline{x} | \ p_j(\underline{x}|A_j) > p_i(\underline{x}|A_i);\ i \neq j\} \qquad (4.5-10)$$

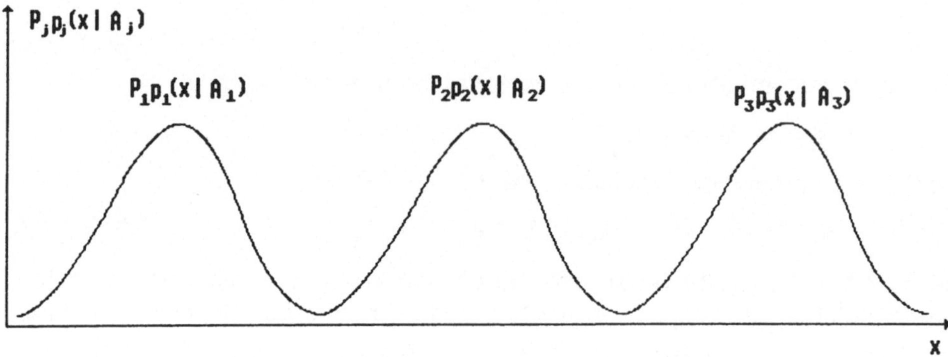

Fig. 4.5–2 Appropriate distribution

If the distributions $p_j(x|A_j)$ are as in fig. 4.5–2, the error probability is low. If, on the other hand, the distributions $p_j(x|A_j)$ are as in fig. 4.5–3, then the resulting error probability is unacceptable for character recognition.

To improve the pre–classification results, it is advisable to define overlapping regions. In fig. 4.5–3, the feature space is divided into a number of subsets $\{I_m\}$. I_2, I_3 and I_4 are essentially overlapping areas. The subsets are assigned by

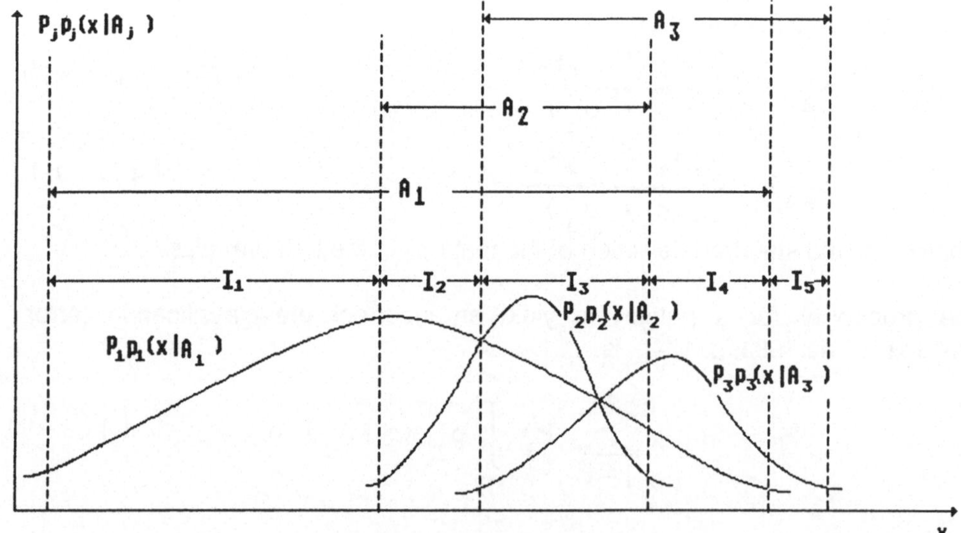

Fig. 4.5–3 Inappropriate distribution

$$I_1 \longrightarrow A_1 \tag{4.5-11}$$

$$I_2 \longrightarrow A_1 \cup A_2 \tag{4.5-12}$$

$$I_3 \longrightarrow A_1 \cup A_2 \cup A_3 \tag{4.5-13}$$

$$I_4 \longrightarrow A_1 \cup A_3 \tag{4.5-14}$$

$$I_5 \longrightarrow A_3 \tag{4.5-15}$$

The subsets $\{I_j\}$ are limited for all pre–classes in the feature space. With a given risk r, the interval (μ_j–x, μ_j+x) for the j–th pre–class with one–dimensional features is computed as

$$\int_{-\infty}^{\mu_j - x} P_j\, p_j(\tilde{x} \mid A_j)\, d\tilde{x} + \int_{\mu_j + x}^{+\infty} P_j\, p_j(\tilde{x} \mid A_j)\, d\tilde{x} \le r \tag{4.5-16}$$

With N–dimensional feature vectors, we receive for the j–th pre–class

$$P_j \int_{B_x} p_j(\underline{x} \mid A_j)\, d\underline{x} \le r, \text{ if } \underline{\mu}_j \in B_\mu \text{ and } B_x = \mathbb{R}^N \setminus B_\mu \tag{4.5-17}$$

The overlap areas are surrounded by the limits of two or more pre–classes. With normal distribution, the determination of the surrounding areas is still possible, albeit tedious. If the densities $p_j(x|A_j)$ cannot be expressed by closed formulas, the determination of the areas surrounding the overlap areas is difficult.

This method will lead to ineffective pre–classification if the instances fall into the overlap area and more than three pre–classes are assigned to this overlap area (see fig. 4.5–4).

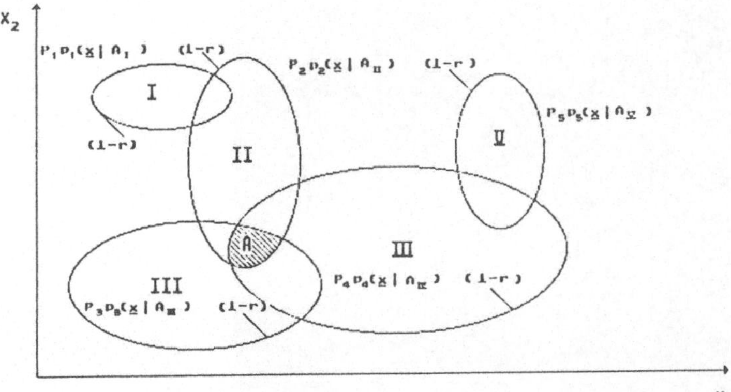

Fig. 4.5–4 Multiple overlaps

4.6 Dynamic Classification

In the dynamic classification as used in the TECHIS project, the subsets for pre-classes are not fixed in advance, but for every instance an individual pre-class is determined dynamically. Overlap problems mostly occur when instances would have to be placed at or near the border of pre-classes, while in dynamic classification, the instance itself is taken to be the centre of its pre-class.

An additional advantage lies in the fact that dynamic classification returns an ordered list of candidates instead of just pointing at a set of characters with equal plausibility. In many instances it is sufficient to consider only the first candidate, thus transcending the limits of a pure pre-classification method. (For example, the experiments over approx. 20,000 Chinese characters with the TECHIS recognition system were conducted with "first-only" classification and still yielded a recognition rate of 98.3 %).

The classification is done as follows: Let x_i^* be the estimated expectation vector for the L feature vectors $\underline{x}_{ij}$ of the i-th class (starting with the reference character L=1)

$$\underline{x}_i^* = [\, x_i^*(1), \ldots, x_i^*(N) \,]^T = \frac{1}{L} \sum_{j=1}^{L} \underline{x}_{ij} \qquad (4.6-1)$$

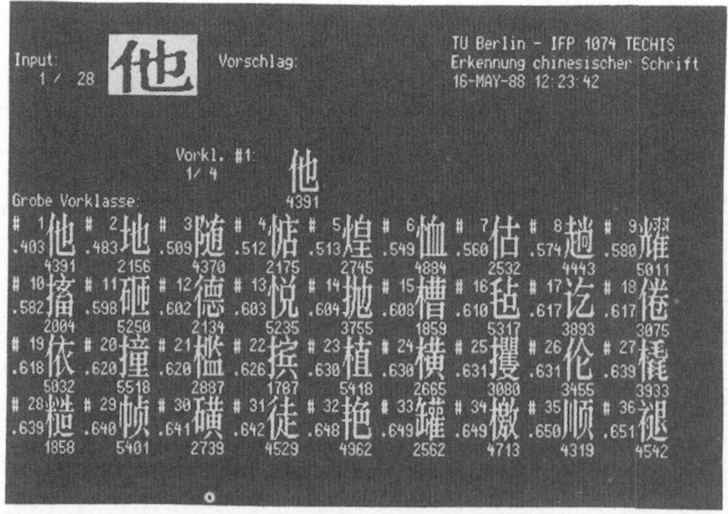

Fig. 4.6–1 Input character with rough pre-class

Let $\underline{x}$ be the feature vector of the unknown character to be recognized, with

$$\underline{x} = [\, x(1), \ldots, x(N)\,]^T \qquad (4.6-2)$$

The pre–class for this character is determined by city block distance. It consists of all $\underline{x}_i^*$ in which the distance to $\underline{x}$ is smaller than a given threshold Δx:

$$PC_x = \left\{ \underline{x}_i^* \,\middle|\, \| \underline{x}_i^* - \underline{x} \|_1 \leq \Delta x \right\} \qquad (4.6-3)$$

The determination of Δx is given by the following relations. Let d_i^* be the average distance in the i–th class, i.e.

$$d_i^* = \frac{1}{L} \sum_{j=1}^{L} \| \underline{x}_i^* - \underline{x}_{ij} \|_1 \;\hat{=}\; \frac{1}{L} \sum_{j=1}^{L} d_{ij} \qquad (4.6-4)$$

The related variances may be estimated from

$$\bar{\sigma}_i^2 = \frac{1}{L} \sum_{j=1}^{L} \left[d_{ij} - d_i^* \right]^2 \qquad (4.6-5)$$

Then Δx has to be chosen as the maximum of all character class deviations, i.e.

$$\Delta x = \max_i \left[\, d_i^* + \alpha \sigma_i \,\right]; \quad \alpha \in [2, 4] \qquad (4.6-6)$$

For 10 extremely frequent characters, a variance analysis was conducted for 100 occurrences each. Average distances ranked from 0.2026 to 0.3280 and seem to be inversely correlated to the complexity of the character (i.e. the higher the complexity, the lower the average distance).

The estimated standard deviations of the average distance ranked between 0.041 and 0.077. A correlation between character complexity and variance could not be noticed.

If the average distances d_i^* show a Gauss distribution, then more than 90 % of all realizations of a given class are found in the set $\{ d_i^* \mp 3 \;\sigma_i \}$. The number of candidates in the pre–class grows with increasing Δx.

4.7 Plausibility Checks

Dynamic classification procedures perform the functions of pre–classification as well as of fine classification. Therefore, it can be used without a separate fine classification by always returning the candidate with least distance to the input character as recognition result. In doing so, a number of **substitutions** (erroneous outputs) will remain, nevertheless. It is up to adequate plausibility checks to correct these errors.

Plausibility checks are defined here as procedures that evaluate the probability of correct candidate selection. **Absolute plausibility checks** test a given distance measure between the candidate and the character currently to be recognized against a threshold value. These threshold values have to be determined either by hand or by adaptive algorithms. In **relative plausibility checks**, the candidate with the extreme (depending to definition, either minimal or maximal) value of a numerical feature is selected from a list of several candidates.

The disadvantage of relative plausibility checks lies in the fact that eventually no unambiguous discrimination is possible, because several candidates share the same extreme value; the advantage is that no threshold value is required. In the TECHIS system, both absolute and relative plausibility checks are used as will be described in the following.

The dynamic pre–classification produces a list of candidates arranged after ascending distance $d(i) := ||\underline{x}_i^* - \underline{x}||_1$:

$$d(i_1) \le d(i_2) \le \ldots.. \le d(i_m) \qquad (4.7-1)$$

where $L_{\underline{x}} = \{i_1,\ldots,i_m\}$ represents the list.

The list length m may be arbitrarily chosen. It should however be taken into account that for small m the probability to include the correct candidate in $L_{\underline{x}}$ decreases. With larger m, on the other hand, more time is required for the calculations.

The candidate list produced in this way is used as a **rough pre–class** for the plausibility check (see fig. 4.6–1 for an example of an input character with its rough pre–class). To further decrease computation time, it is reduced to a **fine pre–class** F as follows:

$$F_{\underline{x}} = \{ i_f | i_f \in L_{\underline{x}} ; d(i_f) \le d_{max} \} \qquad (4.7-2)$$

The threshold d_{max} can again be adjusted to achieve optimal separation under the given conditions. This can also be done by the operator during recognition operation.

If a too low value is chosen for d_{max}, it frequently occurs that

$$d(i_1) > d_{max}, \qquad (4.7-3)$$

i.e. the fine pre–class contains no candidate at all and so is not qualified for plausibility checks. (On the other hand, a low value for d_{max} may well serve as a rejection threshold!)

For these reasons, the definition of the fine pre–class is modified to

$$\tilde{F}_{\underline{x}} = \begin{cases} \{i_f \mid i_f \in L_{\underline{x}} \; ; \; f=1,\ldots,f_{min}\} & \text{if } d_{max} < d\left(i_{f_{min}}\right) \\ \{i_f \mid i_f \in L_{\underline{x}} \; ; \; d(i_f) \le d_{max}\} & \text{else} \end{cases} \qquad (4.7-4)$$

f_{min} denotes the minimum length of the fine pre–class . A f_{min} value of 1 makes sure that the list contains at least one candidate. For relative plausibility checks, $f_{min}>1$ must hold.

4.71 Absolute Plausibility Tests

The simplest absolute plausibility test, the rejection threshold d_{max}, has already been discussed. An appropriate rejection threshold can be determined from experience.

If the pre–classification process delivers not only the ordered list of candidates, but also their respective distance values from the unknown sample, the plausibility of the classification can be further tested with a separation threshold.

Let $\tilde{F}'_x$ be the fine pre–class assigned to the sample $\underline{x}$ and $D_x = \{d_x(i_f)\}$ be the sequence of distances corresponding to the elements of $\tilde{F}_x$, so that $d_x(i_f)$ is the distance between the i_f–th candidate of $\tilde{F}_x$ and $\underline{x}$. The difference distance (separation) of the first and second candidate is expressed as

$$s_{1,2} = d_{\underline{x}}(2) - d_{\underline{x}}(1) \qquad (4.7-5)$$

If the first and second candidate have equal distances to the unknown sample (which may happen when the sample is a representation of an out–of–set character), the difference distance is 0. The plausibility of the first candidate increases with the difference $s_{1,2}$. A threshold $t_{1,2}$ may be used to test the separation:

IF $s_{1,2} > t_{1,2}$
THEN $\underline{x}_{i1}^*$ is a reliable recognition result for $\underline{x}$
ELSE perform further plausibility checks.

Another absolute plausibility test, the **classification cut**, can be found through statistical analysis of distances in dynamic classification. If a feature algorithm delivers appropriate feature values that allow a judgement on whether the correct candidate x_{i1}^* has been reached even before all classes have been tested, an early exit ("cut") of the time–consuming classification loop is possible as soon as

$$d_{\underline{x}}(i_1) < c \qquad (4.7-6)$$

A threshold c has to be determined from a minimal distance in the set of all characters and the appropriate variance analysis. If this threshold is too high, wrong (early) cuts will be produced. If the threshold is too low, on the other hand, characters will not be covered by the cut, and no time is saved. This threshold can be adjusted interactively by the operator to best suit the given recognition conditions.

Every successful cut reduces the computation time required for classification. If the occurrence probabilities of the "cut" characters are equal (or if the candidates are evaluated in an order not correlated with occurrence frequency), an average time saving of 50 % is effected by every cut. If, however, the N candidate classes are ordered according to occurrence frequency, so that the most frequent candidates are evaluated first, the time saved amounts to a much higher percentage.

While an uncut classification cycle demands N evaluations, a character i with frequency p(i) and order rank r(i) on the frequency index list requires r evaluations. Then the average saving is given by

$$\frac{1}{N}\sum_{i=1}^{N}\left[N-r(i)\right]p(i) \qquad (4.7-7)$$

For example, in the TECHIS system the most frequent Chinese character (4.16%) regularly returns a distance well below the the cut threshold. With almost every instance of this character, the classification ends after the first evaluation, saving (3755–1)/3755 = 99.97 % of the classification time required without cut. This one character reduces the overall classification time of the system by 4.15 %. The frequency distribution [Xia 86:1302] allows the prediction that 50 % of running text are covered by 121 characters, and experiments confirmed the resulting tremendous gain in recognition speed.

4.72 Character Frequency

In accordance with the popular concept of plausibility, the candidate with the relatively highest occurrence frequency may be considered the most plausible on a given candidate list. Several lists of occurrence frequencies have been published. In the TECHIS project, the data from [Xia 86] were incorporated into the Chinese characters data base (see 6.3). For every character in the set of characters to be recognized, the occurrence frequency p(i) is stored: for the top 2,400 characters, this is the percent value from [Xia 86]. For the remaining characters it was set to zero.

The frequency check yields the index of the most frequent candidate, provided that only one candidate has the maximum value:

$$h_{\underline{x}} = \begin{cases} i & \text{if } p(i) = \max\left\{p(i_1),\ldots,p(i_N)\right\} \text{ unique} \\ 0 & \text{else} \end{cases} \tag{4.7-8}$$

If there is no dominating candidate, the occurrence frequency is not further analyzed at the present time.

4.73 Pattern Match

The pixel-by-pixel comparison of two matrices (pattern match) is one of the oldest methods in character recognition. In the TECHIS system, pattern match is used as plausibility check to give further discrimination support. Pattern match requires considerable processing time (mostly for input of reference pattern from external storage media, but also for the required size normalization of the character in question). Since it is furthermore sensitive to a number of distortions, it is used only if ambiguities arise.

Let $B=\{b(i;j)\}$ be the binary input character and $R_k=\{r_k(i;j)\}$ be the binary reference character of the k-th candidate in the pre-class ($k \leqslant i_n$), then the pattern match error is given by

$$\delta(B,R_x) = \sum_{i=1}^{n} \sum_{j=1}^{m} |(b(i;j) - r_k(i;j)| \tag{4.7-9}$$

The pattern distance (see 3.31) check yields the index of the candidate i* with minimal pattern difference (error), provided that the minimum value is not shared by several candidates:

$$i^* = \begin{cases} k & \text{if } \delta(B,R_k) = \min\left\{\delta(B,R_1),\ldots,\delta(B,R_{i_n})\right\} \text{ unique} \\ 0 & \text{else} \end{cases} \qquad (4.7\text{-}10)$$

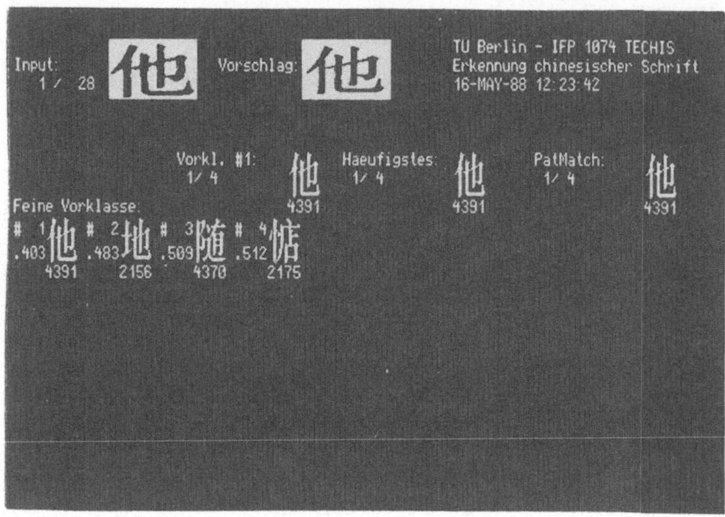

Fig. 4.7–1 Fine pre-class with plausibility checks

4.74 Discrimination Procedure

The interface between pre–classification and plausibility checks is defined by the discrimination procedure which is organized as follows:

1. Determine the fine pre–class $\widetilde{F}_x$;

2. a := i_1;

3. Perform separation test: if $s_{1,2} > t_{1,2}$ then i_1 delivers the recognition result, i.e. $\underline{x}_{i1}{}^*$. Otherwise:

4. Perform frequency test: b := h_X;

5. If a = b, then i_1 delivers the recognition result, i.e. $\underline{x}_{i1}{}^*$. Otherwise:

6. Perform pattern match: c := i^*;

7. If a = c or b = c, then c delivers the recognition result, i.e. $\underline{x}_{i^*}{}^*$. Otherwise (last resort):

8. Interactive menu selection by operator.

This amounts to saying that a candidate is automatically chosen as recognition result if the separation test succeeds, or if two or more of the following conditions hold:

- position 1 in the pre–class;

- most frequent candidate in the fine pre–class;

- least pattern distance from reference pattern.

In all other cases, the selection of the correct result is left to the operator. In practical experiments it was found, however, that the position 1 condition in almost all cases corresponds to the correct result. Full–automatic recognition is feasible with only the position 1 condition (see 5.43 for the results).

4.8 Learning Mechanisms

Occurrences of most Chinese characters are few and far between. Classification methods that need only one sample per class are accordingly required. On the other hand, an optically sampled character pattern is always a cross-modulation of signal and noise. The more samples are evaluated and averaged for one class, the higher is the chance that different "noisy" details in any of the samples is neutralized in the average. Therefore, a learning mechanism can again raise the recognition reliability by no small amount if the size of the sample is larger.

To economize on storage that would otherwise be required, **recursive averaging** was chosen in the TECHIS project. Let $\underline{x}_i^*$ be the estimated expectation vector of the i-th character and M(i) the number of occurrences which are also stored. When the "Learn" mode is active and a sample with feature vector x_{ij} has been recognized successfully, these data are updated as follows:

$$\hat{\underline{x}}_i := \frac{\underline{x}_i^* M(i) + \underline{x}_{ij}}{M(i) + 1} \qquad (4.8-1)$$

$$\underline{x}_i^* := \hat{\underline{x}}_i \quad \text{and} \quad M(i) = M(i) + 1 \qquad (4.8-2)$$

In similar fashion, the variance $\sigma_i^2(k)$ of the k-th vector component x_i can be determined recursively. Let

$$\underline{x}_i^* = [\, x_i^*(1), \ldots, x_i^*(N)\,]^T \qquad (4.8-3)$$

be the estimated expectation after M(i) samples and

$$\underline{x}_{ij} = [\, x_{ij}(1), \ldots, x_{ij}(N)\,]^T \qquad (4.8-4)$$

be the current sample for the i-th character, then the variance is updated by

$$q := M(i) + 1 \qquad (4.8-5)$$

$$\hat{\sigma}_i^2(k) := \frac{M(i)\,\sigma_i^2(k) + \left[x_i^*(k) - x_{iq}(k)\right]^2}{q}, \quad k = 1, \ldots, N$$

$$(4.8-6)$$

$$\sigma_i^2(k) := \hat{\sigma}_i^2(k); \quad M(i) =: q. \qquad (4.8-7)$$

The occurrence counters M(i) are initialized with the value 1 for the reference character. The first sample contributes 50 % to the resulting feature vector $\hat{\underline{x}}_i$, the second 33 %, and so on. After a certain number of samples, the effect of learning diminishes to a degree that it might as well be turned off again (but only for the current character class!). This can be tested by fixing a threshold c_i and performing the feature update only if

$$|| \underline{x}_i^* - \hat{\underline{x}}_i ||_1 > c_i , \qquad (4.8-8)$$

but to compute the distance produced by the learning, the learning itself must have been done before. The only time saved by the c_i test is for copying the updated feature vector $\hat{\underline{x}}_i$ to the location of $\underline{x}_i^*$, which takes almost no time when done in memory. Therefore, at present we renounce an c_i test.

Other learning mechanisms may be envisioned. Bit-packed binary reference patterns resist to recursive averaging, though. They can only be averaged if a number of samples is accessible at the same time: the pixel values are added for every element position, and the resulting pseudo-grayscale image is binarized again with a suitable threshold.

The analysis of the local character and word frequencies in a given text promises to produce a helpful plausibility test. This can however not be considered a learning mechanism, because the value of local frequencies lies in their locality. The more averaging is done here, the more information is lost. Finally, the counting of character or word transition occurrences is another possible application for a learning mechanism.

5 The TECHIS System: Implementation and Results

5.1 Hardware and Software Conditions

The hardware used in realizing the TECHIS recognition system consists of a number of devices. The following computer systems were used:

a) VAX 11/780 under the Ultrix-II operating system (UNIX-compatible) for the character database (see 5.3) and a number of feature tests;

b) MicroVAX II under the VMS operating system with a connected VTE MicroPicture 200 picture processing system (resolution 512*512 pixels at 256 gray levels) for development and research;

c) Atari Mega ST4 under GEM-TOS with a connected PrintTechnik PRO 8805 video digitizer (resolution 256*512 pixels at a maximum of 127 gray levels) for demonstrations;

d) IBM PC-AT under MS-DOS/Windows for scanner operation.

Data transmission between the computers was conducted via RS232 ports. In the case of a) and b), data was transferred over the university telephone net with acoustic couplers.

Input of Chinese characters was done with a Siemens K211 CCD video camera linked either to the picture processing system at b) or the video digitizer at` c). Input control was achieved by feeding the video signal through an analog monitor in c), while the picture processing system at b) allowed also analog video output.

A flat-bed CCD scanner (Siemens ST 400, pre-production sample) was used for a limited time to scan Chinese texts into d). The data was then transmitted to b) over RS232 ports which took about half an hour per page. Despite of this limitation, the fundamental feasability of scanner input could be verified.

Output of Chinese characters was done on screen and paper. For "soft copy" (screen output), subroutines were written that display characters from 16*16 pixels to 64*64 pixels on bit-mapped screens (the video monitor at the picture processing system resp. the Atari monitor). Dedicated print programs were developed on b) for driving 16-pin needle and 24-jet ink-jet printers. They take

ASCII-coded GB codes as input and produce 16*16 resp. 24*24 matrix output from font files included in the database.

Most programs were written in the FORTRAN-77 programming language. On c), some 68000 assembly language subroutines were necessary for screen and digitizer control. The programs for the database system on a) were written in Pascal.

5.2 Program Implementation

A number of feature and classification algorithms was implemented and tested in the TECHIS project. Best results were found with the BJD-BS feature in one experiment and a combination of ET1, ET2, SP4 and 4-SDF features in another.

The BJD-BS (black jump distribution in balanced subvectors) feature with dynamic classification was used in implementing two versions of the recognition system both on the MicroVax and the Atari computers. Both consist of programs for the input and preprocessing, segmentation, feature extraction, classification and output phases. Additional utility programs perform feature generation and evaluation, as well as filtering and editing the 64*64 reference character patterns.

On the MicroVax, the general development version is used, while for demonstrations a special version was re-created on the Atari. Because of its limited speed, compromises had to be taken that however also brought valuable results.

Size normalization in the Atari version was done with linear interpolation, in contrast to the two-dimensional spline interpolation on the MicroVax which took more than a minute per character on the Atari. After linear interpolation, the normalized characters appeared rounded at the edges and slighly out of focus, but recognition results were nevertheless comparable to the MicroVax system (see below). If the raw input characters occupy a rectangular area smaller than 64*64 pixels, size normalization can be turned off altogether which again increases speed but does not notably degrade the recognition rate.

In the Atari system, the classification cut was also introduced as a means of increasing the recognition speed. It proved extremely useful. With an appropriate cut threshold at about 0.6 (depending on the text conditions), most characters were covered by the cut while "early cuts" could almost totally be avoided. The only frequent characters that often yielded feature values well above the cut threshold were "yi (one)" and "ge (numerator)". For these characters, the complete classification run over 3,755 characters in Level 1 of the GB 2312-80 still had to be performed.

5.3 A Database on Chinese Characters

A database was created for collecting the available informations on Chinese characters (pattern bit-maps in various sizes, character frequency data, as well as linguistic data on characters, words and phrases). The organisation of most parts of the database is character-oriented. The GB 2312-80 code serves as main key for each Chinese character. See fig. 5.3-1 for the structure of the database.

5.31 Reference Patterns

For all characters in the working set (currently level 1 of the GB 2312-80, i.e. the most frequent 3,755 characters in PRC simplification), a reference pattern in *Songti* style was recorded by CCD camera, a printed copy of the GB 2312-80 serving as original. The characters were normalized in size to 64 by 64 pixels and binarized with global threshold on the transformed picture (see 2.4). The

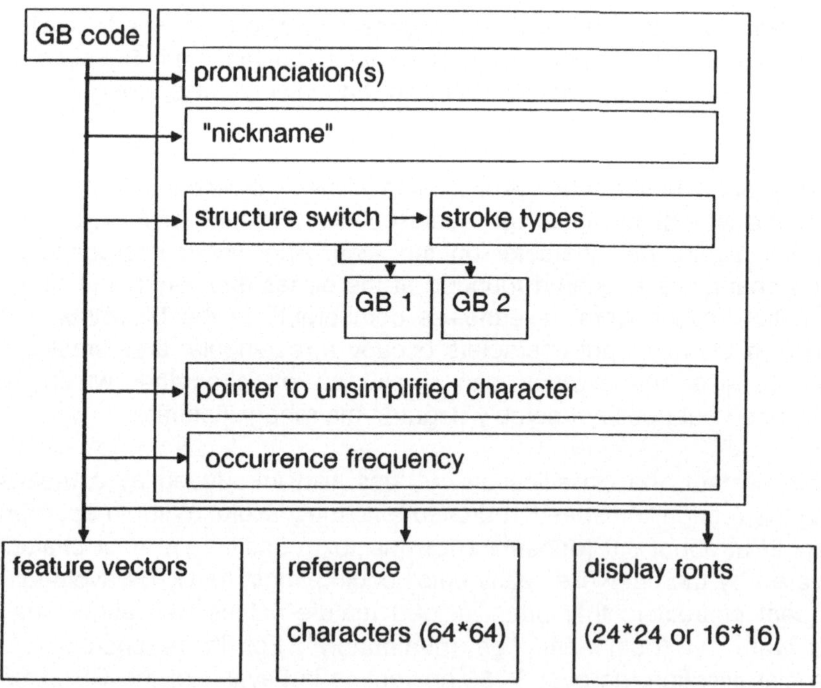

Fig. 5.3-1 Database structure

matrices are densely packed, a bit in the data record representing a pixel in the binary picture, and stored in a random access file. The record length required is 512 bytes plus the length of the record header. The contents of this record header are set to zero for reference pattern files. For other pattern files, it contains the GB 2312–80 codes of the current record and a pointer at the next character in logical order.

A second version of the reference pattern file was produced by applying Zhang's smoothing algorithm (see 2.5) to the individual characters. The format is compatible with the original reference pattern file. Experiments showed that from this file version, more stable feature values could be generated.

The reference pattern file is used in the following processes:

1) Creation of a feature file: The feature computation algorithm is run sequentially on all characters in the reference pattern file chosen, the output being written in the same order to a random–access feature file. This is a lengthy process that requires several hours, depending on the feature algorithm being used, but can be run unattended.

2) Pattern match: When the classification algorithm cannot select a candidate with sufficient reliability, an additional pattern match may be performed in which the minimum pattern distance is determined between the current input pattern and the reference patterns of the candidates in the pre–class that are loaded into memory from the reference pattern file.

3) Screen output: After successful completion of the recognition process, the reference pattern of the result character is displayed on–screen as a 64 by 64 pixel matrix, allowing comparison with the input character. Since the reference characters still contain distortions, they are not very suitable for high–resolution printer output. For this purpose, more intelligent smoothing algorithms have to be developed, while the ultimate pixel editing has to be done manually.

In the near future, reference pattern files will also be created for other fonts (*fangsongti, heiti, kaiti*) to allow comprehensive tests on font differences.

5.32 Frequency Data

Character frequencies are computed as Nx/N, where Nx is the number of occurrences of the character x and N is the total number of character occurrences (tokens) in the underlying set of texts (corpus).

From the number of frequency counts undertaken so far, we chose the one done by the Beijing Languages Institute and the Language and Script Committee

[Xia 86], where N = 1,771,398. This list contains frequency values for 4,574 distinct character types. We ignored characters with an absolute occurrence count of 1 or 2. Of the remaining 3,889 characters, 2,592 or 66.65 % were found in Level 1 of the GB 2312–80, meaning that about one third of the frequency–ranked characters are in Level 2 or even out of set. On the other hand, Level 1 also includes 1,163 characters or 30.97 % that were not included in the truncated frequency list.

The frequency data are stored as real numbers and form a part of the combined database file. For GB characters not in the truncated frequency list, the frequency was set to zero.

The frequency information was used to create a frequency–ordered index list that performs a frequent–first search and thus increases classification speed when the Cut mechanism is enabled. This list is also stored as a sequential binary file. It needs only to be recomputed if the frequency data change. In the recognition process, the frequency data are also evaluated as one plausibility check (see 4.32).

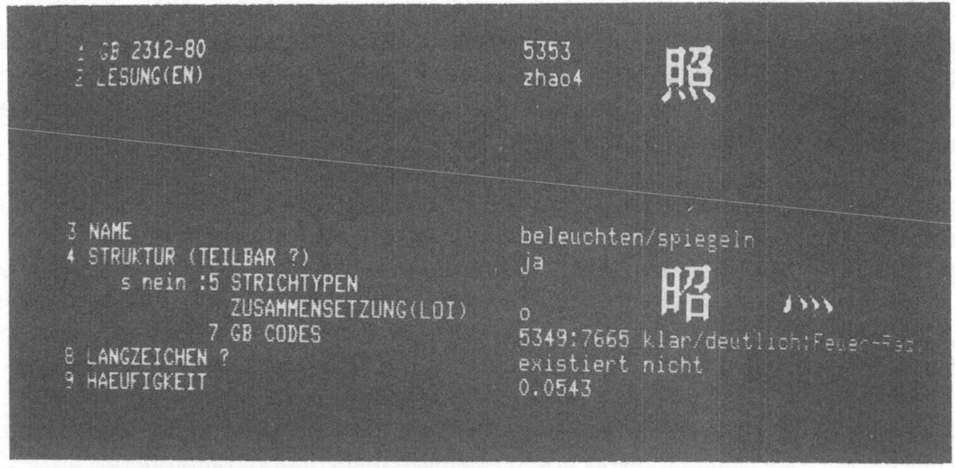

Fig. 5.3–2 Screen display of the database program

5.33 Linguistic Data: Characters

For future applications of the TECHIS system, a number of linguistic data was also collected for all characters in the GB 2312–80 (of which the current working set is a subset, Level 1) and some extensions. Input and editing of this part of the data-base was done with a dedicated program that allows graphic output on VT240 terminals (see fig. 5.3–2 for a display hardcopy).

The character-oriented data include:

– up to 6 pronunciation(s) of the character as specified in the Pinyin index of the GB 2312–80. The tone is marked by a numeral behind the Pinyin syllable. If a character has less than 6 readings, the remaining fields are filled with blanks. If a component has no reading at all, all fields are blank-filled;

– a "nickname" of max. 20 ASCII-characters, giving a hint at the meaning of the character, in German;

– structure data:
– a Boolean switch to indicate whether the character is a binary compound;

– for non-compound characters, a code string for the strokes of which it is composed in conventional stroke order;

– for compound characters, a marker for the composition type (possible types are left/right, top/bottom, and inside/outside which also serves as a catch-all rest class for rare composition types) and pointers at the components.

If the components are valid characters included in the GB 2312–80 (this occurs with 2/3 of the characters), their GB number (in the range 1601–8794) is used as pointer. In other cases, the pointer refers to an "extended GB" code locally defined in the TECHIS project.

Out-of set characters were first tested for inclusion in the *Cihai* dictionary [Cih 79]. 123 characters for which the *Cihai* gives at least one reading were considered valid characters and assigned a code number above 8800. 599 components for which no reading exists were coded in the range 101–1594.

In coding the New Chinese-German Dictionary, a research team (Heinzl/Ding/Lachner) at the Technical University of Munich found 241 characters used in this dictionary but undefined in the GB 2312–80. These characters were assigned extended GB codes of above 9000. To ensure data compatibility, the TECHIS database uses the same code numbers for these characters.

By storing the structure data as binary composition type and pointers to components (which themselves may have corresponding structure data), the decomposition of characters into primitive components allows the creation of binary structure trees (see fig. 5.3–3).

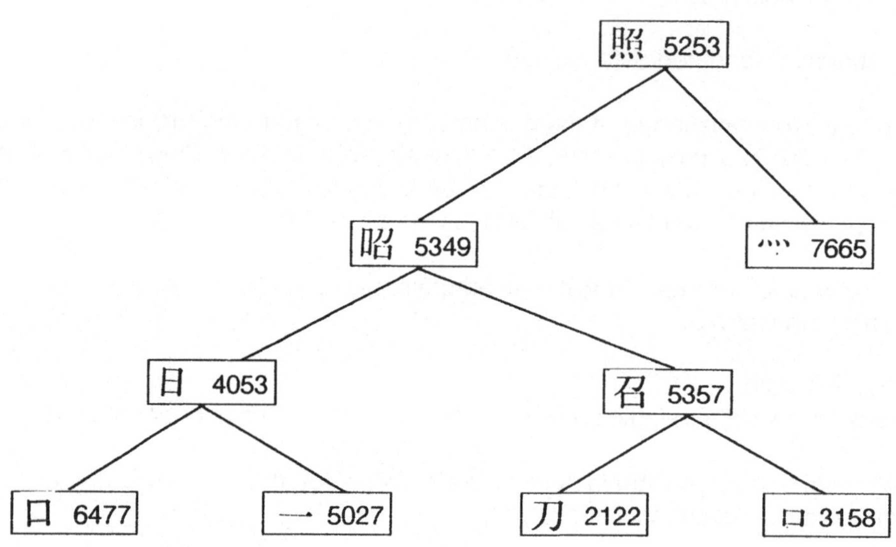

Fig. 5.3–3 Binary tree representation of character structure

5.34 Linguistic Data: Words and Phrases

For use in further research, a file type for word and phrase lists has been defined. Files of this type are to contain the GB codes of the characters making up the word/phrase, the pinyin spelling of the word/phrase, a "nickname" in German and a frequency value which will also be copied from the related lists in [Xia 86].

Files of this type will be used as a further plausibility check to analyze the character–oriented output of the recognition system. The inclusion of "nicknames" makes it also possible to perform a word–by–word transposition, which is not a translation, of the Chinese text to German. Different files will be created for different categories of text, and one area of research will be the automatic determination of text categories and thus automatic selection of the appropriate word/phrase file.

5.4 Tests on Features

5.41 Feature Files

Feature files contain the feature values for all characters in the working set. They are created as described above by running the feature algorithm on all working set characters and storing the results in the same order into a random-access file (see 5.31). The record length depends on the size of the feature vector. In the BJD-BS algorithm, this amounts to 32 real values (plus an integer giving the absolute number of occurrences so far, required by the learning mechanism). In the Z2 algorithm, 128 byte values are used.

The feature files are used by the classification algorithm. At program start (or whenever the user requires it) the complete feature file is read into memory to speed up the classification process. If learning (recursive averaging) acts have taken place, the updated feature file is written back to disk at program end or on demand.

Approximate feature files for a given font can easily be generated by loading a default feature file (in which all occurrence counts equal 1) and updating all feature vectors for occurring characters in that font. Afterwards, the feature space is saved under a new filename that indicates the font name. In subsequent learning sessions, this feature file is loaded and updated again. Characters that have not yet been "learned" in the specific font are still treated according to their default values.

Currently, approximate feature files exist for BJD-BS features of *heiti* and *kaiti* fonts as well as for non-simplified *songti* characters. In one experiment particularly worth mentioning, part of the default feature file (GB 1649..81) was updated with 66 Russian (cyrillic) upper- and lowercase letters taken from the Pravda. After typically 4 to 5 learning samples per letter (which were actually averaged with the underlying original Chinese character), all letters could be recognized correctly – although every recognition pass evaluated the full 3,755 character set!

5.42 Test Data

For a comprehensive test of various features, 15 pages of the Chinese magazine "Xiandaihua" (Modernization) were read in by camera. The characters were normalized in size to 64*64 pixels, binarized and stored in a separate file for each page. This set of files, designated CTX1 (Chinese texts 1), has the following contents:

File No.	Char.s	Chinese char.s
4	588	565
5	1,093	1,057
6	1,481	1,299
7	1,702	1,447
8	1,661	1,463
9	1,706	1,528
10	1,555	1,367
11	1,806	1,624
12	1,447	1,297
13	1,369	1,235
14	1,721	1,531
15	879	746
17	1,640	1,453
18	1,553	1,374
19	1,417	1,251
Total	21,618	19,241

Of these 19,241 characters, the overwhelming majority (19,231 or 99.95 % instances of 1,544 different characters) were found to be included in Level 1 of the GB 2312–80 and thus eligible for our recognition experiments. Notice that in spite of the considerable sample size, less than half of the 3,755 characters in Level 1 actually occurred. For the most frequent character, 770 instances were counted. Of the remaining ten characters, one was a non–abbreviated one (although this should not happen in a modern PRC text), and nine came from Level 2 of the GB 2312–80 (see fig. 5.4–1).

In order to test the "re–recognition" of the reference characters, most of the reference characters were read in one more time by a 400 dpi flat–bed scanner. This file GB2 contains 3,413 different Chinese characters.

啬 旎 锲 崛 髦 麟 匾 並

Fig. 5.4–1 Level 2 characters in the CTX1 sample

5.43 Results for the BJD–BS Feature

The BJD–BS (black jump distribution in balanced subvectors) feature was tested against both the CTX1 and GB2 character files. The classification was simplified to just picking the candidate with the least linear distance (position 1 criterion) as recognition result. A reject threshold was employed, but no plausibility checks were used. The recognition rates (number of correctly recognized characters, divided by the number of Chinese characters in text) for CTX1 ranked from 96.1% to 99.1 % for individual pages, the total recognition rate was 98.21 %.

File	No.Char.s	Chinese char.s	correct	substitutions	rejections	rec.rate
4	588	565	559	2	4	98.94
5	1,093	1,057	1,047	7	3	99.05
6	1,481	1,299	1,269	25	5	97.69
7	1,702	1,447	1,390	44	13	96.06
8	1,661	1,463	1,445	13	5	98.77
9	1,706	1,528	1,502	12	14	98.30
10	1,555	1,367	1,337	24	6	97.81
11	1,806	1,624	1,577	31	16	97.11
12	1,447	1,297	1,270	19	8	97.92
13	1,369	1,235	1,222	10	3	98.95
14	1,721	1,535	1,517	15	3	98.83
15	879	746	734	12	0	98.39
17	1,640	1,453	1,435	11	7	98.76
18	1,553	1,374	1,355	14	5	98.62
19	1,417	1,251	1,237	14	0	98.88
Total	21,618	19,241	18,896	230	88	98.21

The re–recognition experiment on GB2 yielded 99.5 % (3,396 out of 3,413 different characters). Average recognition time was 2.85 seconds per character.

For a number of frequent characters, the estimated expectation values and standard deviations for the distance $||\underline{x}_k - \overline{\underline{x}}_k||$ between 100 actual instances each and their average were calculated (see fig. 5.4–2), showing a remarkable stability. As a further example for separation between different characters, the average distance $||\underline{x}_j - \underline{x}_k||$ and standard deviation are shown for the frequent character 了 and the three other characters with least relative distance:

j	m	s	char.	samples	
1641	0.32803	0.07484	了	100	(j=k)
3717	0.93498	0.07582	子	25	
850	0.96619	0.03262	工	2	
3348	1.04629	0.05325	于	5	

不 (k = 215)		
m :	0.2392692	0.23 ∓ 0.05
s :	5.4510750E-02	
这 (k = 3544)		
m :	0.2025501	0.20 ∓ 0.07
s :	6.9270566E-02	
有 (k = 3338)		
m :	0.2529060	0.25 ∓ 0.05
s :	5.1867731E-02	
在 (k = 3442)		
m :	0.2389332	0.24 ∓ 0.04
s :	4.0526718E-02	
个 (k = 838)		
m :	0.2606088	0.26 ∓ 0.05
s :	5.5236567E-02	
是 (k = 2483)		
m :	0.2277880	0.23 ∓ 0.05
s :	5.0273925E-02	
了 (k = 1641)		
m :	0.3280316	0.33 ∓ 0.08
s :	7.7484146E-02	
的 (k = 506)		
m :	0.2310383	0.23 ∓ 0.05
s :	5.1073954E-02	
发 (k = 660)		
m :	0.2077489	0.21 ∓ 0.04
s :	4.4407751E-02	

Fig. 5.4–2 Estimated expectation values and standard deviations

5.44 Results of Feature Combination

The feature combination ET1, ET2, SP4 and 4-SDF (distance from edge to first resp. second black jump; stroke proportion; stroke density) was in another experiment tested against the same data. For CTX1, the total recognition rate amounted to 99.92 %. Among 19,241 valid Chinese characters, only seven substitutions and nine wrong rejects occurred, mostly with degraded character samples (see fig. 5.4-3):

File	No.Char.s	Chinese char.s	correct	substitutions	rejections	rec.rate
4	588	565	565	0	0	100.00
5	1,093	1,057	1,057	0	0	100.00
6	1,481	1,299	1295	1	3	99.69
7	1,702	1,447	1,443	2	2	99.72
8	1,661	1,463	1,463	0	0	100.00
9	1,706	1,528	1,528	0	0	100.00
10	1,555	1,367	1,364	1	2	99.78
11	1,806	1,624	1,622	1	1	99.88
12	1,447	1,297	1,297	0	0	100.00
13	1,369	1,235	1,235	0	0	100.00
14	1,721	1,535	1,535	0	0	100.00
15	879	746	746	0	0	100.00
17	1,640	1,453	1,453	0	0	100.00
18	1,553	1,374	1,372	1	1	99.85
19	1,417	1,251	1,250	1	0	99.92
Total	21,618	19,241	19,225	7	9	99.92

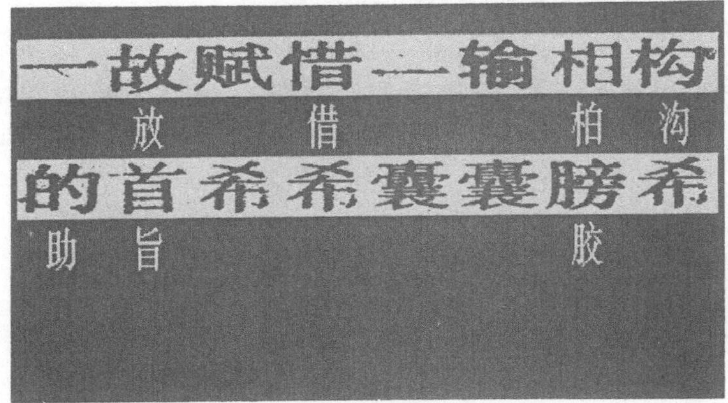

Fig. 5.4–3 Substitutions and rejections

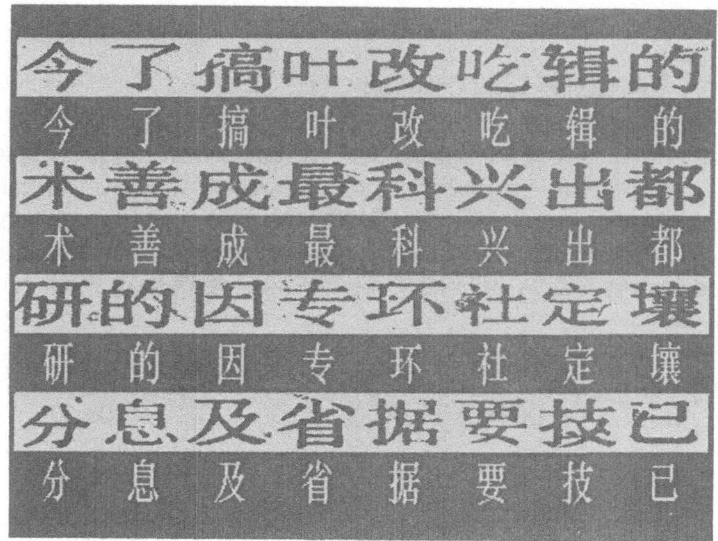

Fig. 5.4–4 Degraded characters correctly recognized

All non–Chinese characters (alphanumerics and punctuation marks) were correctly rejected. The recognition time required was 3.2 seconds per character.

The feature combination described has shown an outstanding degree of reliability against various disturbances caused by low print quality (see fig. 5.4–4)

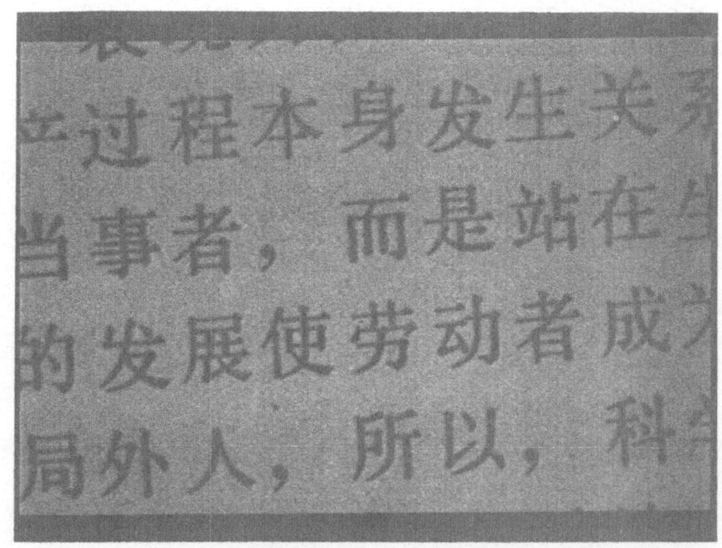

Fig. 5.4–5 Tilted characters (camera picture)

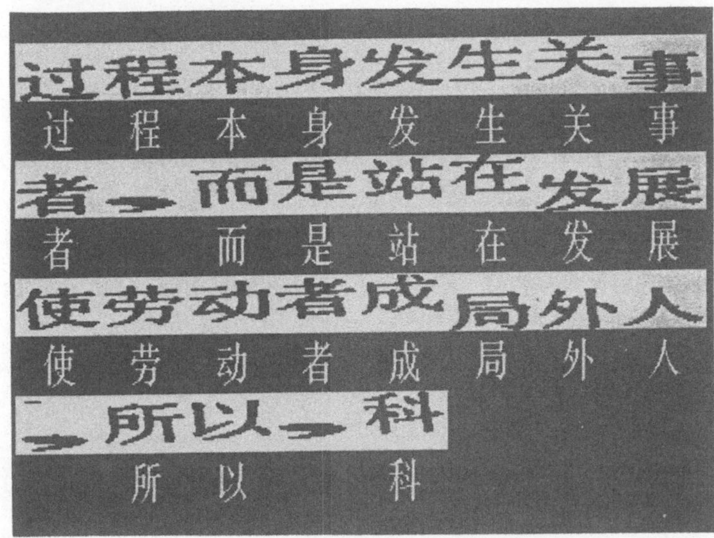

Fig. 5.4–6 Recognition of tilted characters

and seems to be one of the most promising approaches for printed Chinese character recognition. Fig. 5.4–5 and 5.4–6 show a slightly rotated image of Chinese text and its recognition result. Note that all characters were recognized correctly in spite of the rotation.

References

[Bar 68] Bartz, M.R.: The IBM 1975 Optical Page Reader. Part II: Video thresholding system. In: IBM J. of Research and Development 12.5 (Sept. 68) 354–363

[Bec 85] Becker, Joseph D.: Typing Chinese, Japanese, and Korean. In: Computer (IEEE Comp. Soc.) Jan. 1985, 27–34

[Cas 66] Casey, R.; Nagy, G.: Recognition of printed Chinese characters. In: Trans. IEEE (EC–15) 91–101

[Cih 79] Cihai. ("Sea of Words", a large dictionary). Beijing, Shanghai: Shangwu Yinshuguan (The Commercial Press) 1979

[Daw 87] Dawson, Benjamin M.: Introduction to Image Processing Algorithms. In: Byte 12.3 (March 1987) 169–186

[DeF 84] DeFrancis, John: The Chinese Language. Fact and Fantasy. Honolulu: University of Hawaii Press 1984

[Fu 82] Fu, K.S.: Syntactic Pattern Recognition and Applications. Prentice–Hall, Englewood Cliffs, N.J. 1982

[G+X 86] Guo, Chunbiao; Xuan, Guorong: Automatic Recognition of printed Chinese Characters by Four Corner Code. In: 8th International Conference on Pattern Recognition, Paris, Oct. 1986 (IEEE Proceedings) 1013ff

[G+Z 87] Guo, Bao–Lan; Zhang, Cai–Lu: Point Tracking Inclusive Matching Method with Backtracking Control Strategies. In: Proceedings of 1987 International Conference on Chinese Information Processing (vol. II) 309ff

[Gro 67] Groner, Gabriel F.; Heafner, J.F.; Robinson, T.W.: On–line computer classification of handprinted Chinese characters as a translation aid. In: IEEE Trans. EC–16 (Dec. 1967) 856–860

[Hag 81] Hagita, Norihiro; Masuda, Isao: Recognition of handprinted Chinese characters by directions of strokes. (In Japanese, English abstract). Papers of IECEJ Technical Group on Pattern Recognition and Learning, PRL 81–13

[He 87] He, Houcun; Sheng, Huanye: 1987 Automatical Stroke Analysis of Chinese Characters. In: Proceedings of 1987 International Conference on Chinese Information Processing, vol. II, 339–343

[Hir 80] Hirai, Shouichi; Sakai, Kunio: Development of a High Performance Chinese Character Reader. In: Proc. 5th Int'l Conf. on Pattern Recognition (1980) 867–871

[HK 84] Chinese–character processing for computerized bibliographic information exchange: summary report of an international workshop held in Hong Kong, 17–20 December 1984. Ottawa, Ont., 1985. (Proceedings series/IDRC).

[Hua 85] Huang, Jack Kai–tung: The input and output of Chinese and Japanese characters. In: Computer (IEEE Comp. Soc.) Jan. 1985, 18–24

[HUM 79] Hagita, Norihito; Umeda, Michio; Masuda, Isao: Classification of hand–printed Chinese characters by stroke density and two other features. (In Japanese, English abstract). Papers of IECEJ Technical Group on Pattern Recognition and Learning, PRL 79–27

[Jen 86] Jeng, Bor–Shenn; Lin, Chih–Heng: Chinese Character Segmentation Using the Character–Gap Feature. In: Applications of Digital Image Processing IX, vol.697 Aug. 1986, San Diego, California, USA, SPIE Proceedings, 262ff

[Jun 87] Jun, S.; Chung, Ma–Lung: Separating Similar Complex Chinese Characters by Walsh–Transform. In: Pattern Recognition 20.4 (1987) 425ff

[Kar 62] Karlgren, Bernhard: Sound and Symbol in Chinese. Revised ed. Oxford/Hong Kong University Press 1962

[Kim 84] Kimura, F.; Harada, T.; Tsuruoka, S.; Miyake, Y.: Modified Quadratic Discriminant Functions and the Application to Chinese Character Recognition. In: Proceedings of Seventh International Conference on Pattern Recognition, Washington D.C., 1984. (Vol.1) 377ff

[Kit 86] Kittler, J.: Feature Selection and Extraction. In: [Young 86: 59–83]

[Kra 68] Kratochvil, Paul: The Chinese Language Today. London: Hutchinson University Library 1968, reprinted 1970

[L+S 83] Lackschewitz, C.; Suchenwirth, R. (ed.): A Bibliography on Computer Applications in Chinese Language and Script Processing. Hamburg: Asia Documentation Center 1983

[Li 86] Li, Bin; Zhao, Shuxiang: A New Approach to Recognition of Both Handwritten and Multi–Font Printed Chinese Characters. In: 8th International Conference on Pattern Recognition, Paris, Oct. 1986 (IEEE Proceedings) 641ff

[Lia 87] Liang, Gang; Wu, Yu-Pu: A Recognition System of Printed Chinese Text. In: Proceedings of 1987 International Conference on Chinese Information Processing (vol. II) 294ff

[Liu 84] Liu Yu; Kasvand, T.: A New Approach To Machine Recognition of Chinese Characters. In: Proceedings of Seventh International Conference on Pattern Recognition, Washington D.C. 1984 (vol.1) 381–384

[Lu 87] Lu, Haoru: Attributed Grammar for Description of Chinese Characters. In: Proceedings of 1987 International Conference on Chinese Information Processing (vol. II) 344ff

[Mae 82] Maeda, K.; Kurosawa, Y.; Asada, H.; Sakai, K.; Watanabe, S.: Hanprinted KANJI Recognition by Pattern Matching Method. In: Proceedings of Sixth International Conference on Pattern Recognition, Washington D.C., 1982.(Vol.2) 789ff

[Ma 84] Ma Yung-Lung; Jour Jung-Shin: New Probabilistic Model for Chinese Character Recognition. In: Proceedings of Seventh International Conference on Pattern Recognition, Washington D.C., 1984 (Vol.1) 370ff

[Mak 85] Makino, Hiroshi: Beta: An Automatic Kana-Kanji Translation System. In: Computer (IEEE Comp. Soc.) Jan. 1985, 46–52

[Man 86] Mantas, J.: An Overview of Character Recognition Methodologies. In: Pattern Recognition, Pergamon Press, Oxford 1986 (Vol.19, No.6) 425ff

[Mat 85] Matsuda, Ryouichi: Processing Information in Japanese. In: Computer (IEEE Comp. Soc.) Jan. 1985, 37–45

[McC 87] McCormick, John: Text Scanners for the IBM PC. In: Byte 12.4 (April 1987) 233–238

[Nag 80] Nagao, M.: Data compression of Chinese character patterns. In: Proc. of the IEEE 68, 7 (1980) 818–829

[Nak 72] Nakata, K.; Uchikura, Y.: Recognition of Printed Chinese Characters. In: Proc. Conf. Machine Perception of Patterns and Pictures (Teddington, England, April 1972), 45–52

[Nak 73] Nakano, Y.; Nakata, K.; Uchikura, Y.; Nakajima, A.: Improvement of Chinese Character Recognition Using Projection Profiles. In: Proc. Joint Conf. Pattern Recognition 1 (1973) 172–178

[Nak 82] Nakagawa, M.; Aoki, K.; Manabe, T.; Kimura, S.; Takahashi, N.: On-Line Recognition of Handwritten Japanese Characters in JOLIS-1. In: Proceedings of Sixth International Conference on Pattern Recognition, Washington D.C. 1982 (Vol.2) 776ff

[Nie 74] Niemann, H.: Methoden der Mustererkennung. Frankfurt (M): Akad. Verlagsges. 1974 (Informationsverarbeitung in technischen, biologischen und ökonomischen Systemen. 2.)

[Ots 79] Otsu, N.: A threshold selection method from gray-level histogram. In: IEEE Trans. SMC-9 (Jan. 1979) 62-66

[Pan 86] Pan, Bao-Chang: Floating Mask Method For Extracting Hand-Printed Character Features. In: Proceedings of Eighth International Conference on Pattern Recognition, Washington D.C. (1986) 324ff

[Pav 77] Pavlidis, T.: Structural Pattern Recognition. Berlin: Springer 1977

[Ran 65] Rankin, Bunyan Kirk III: A Linguistic Study of the Formation of Chinese Characters. Dissertation, University of Pennsylvania 1965

[Ric 84] Richetin, M.; Vernadat, F.: Efficient Regular Grammatical Inference for Pattern Recognition. In: Pattern Recognition, Pergamon Press, Oxford 1984 (Vol.17, No.2) 245ff

[Ros 82] Rosenfeld, A.; Kak, A.C.: Digital Picture Processing. Academic Press, New York, 2nd ed. 1982

[Rub 88] Rubinstein, Richard: Digital Typography. An Introduction to Type and Composition for Computer System Design. Reading, MA, etc.: Addison Wesley 1988

[She 85] Sheng, Jian: A Pinyin Keyboard for Inputting Chinese Characters. In: Computer (IEEE Comp. Soc.) Jan. 1985, 60-63

[Shr 84] Shridhar, M.; Badreldin, A.: High Accuracy Character Recognition Algorithm Using Fourier and Topological Descriptions. In: Pattern Recognition 17.5 (1984) 515ff

[Shr 86] Shridhar, M.; Baldredin, A.: Recognition of Isolated and Simply Connected Handwritten Numerals. In: Pattern Recognition 1986 (Vol. 19, 6) 1ff

[Sta 76] Stallings, William: Approaches to Chinese Character Recognition. In: Pattern Recognition 1976 (Vol. 8) 87-98

[Suen 86] Suen, Ching Y.: Character Recognition by Computer and Applications. In: [Young 86: 569–586]

[Tam 78] Tamura, H.: A comparison of line thinning algorithms from digital geometry viewpoint. In: Proc. 4th Int. Joint Conference on Pattern Recognition (1978) 715–719

[Tsu 84] Tsukumo, J.; Asai, K.: Non-linear Matching Method for Handprinted Character Recognition. In: Proceedings of Seventh International Conference on Pattern Recognition, Washington D.C., 1984 (Vol.2) 770ff

[Tsu 86] Tsukumo, Jun; Asai, Ko: Machine printed Chinese and Japanese character recognition methods and experiments for reading Japanese pocket books. In: Proceedings CVPR '86, Washington, DC, USA (IEEE Comp. Soc. Press 1986) 162ff

[Tsu 87] Tsukumo, Jun; Asai, Ko: Experiments for Reading Japanese Pocket Books. In: NEC Res. & Dev. 85 (April 1987) 60ff

[Tuc 82] Tuceryan, M.; Ahuja, N.: Segmentation of Dot Patterns Containing Homogenous Clusters. In: Proceedings of Sixth International Conference on Pattern Recognition, Washington D.C. 1982 (Vol.1) 392ff

[Ume 79] Umeda, Michio: Pre–Classification for Recognition of Multi–Font Printed Chinese Characters. (In Japanese). In: Proc. IECEJ J62–D.11 (1979) 758–765

[Ume 82] Umeda, Michio: Recognition of Multi–Font Printed Chinese Characters. In: Proceedings of Sixth International Conference on Pattern Recognition, Washington D.C. 1982 (Vol.2) 793–796

[Ume 84] Umeda, Michio: A multi–font printed Chinese character reader. In: Trans. IECEJ J64–D.8 (1984) 908–915

[Wei 61] Wei Juxian: Wenzixue jiangyi. (Lecture notes on character studies). Hong Kong: Guangxia Shuyuan 1961

[Wid 88] Widmer, Urs: Chinesisch in Bits und Bytes. In: Das neue China 15.3 (1988) 36–39

[Wie 27] Wieger, Leon: Chinese Characters. Their Origin, Etymology, History, Classification and Signification. New York: Paragon/Dover 1965 (Reprint of the 2nd edition of 1927)

[Woo 85] Wood, H.; Reifer, D.J.; Sloan, M.: A Tour of Computing Facilities in China. In: Computer (IEEE Comp. Soc.) Jan. 1985, 80–87

[Xia 86] Xiandai Hanyu Pinlü Cidian. (Frequency dictionary of the modern Chinese language). Beijing: Yuyan Xueyuan Chubanshe (Languages Institute Press) 1986

[Xie 88] Xie, S.L.; Suk, M.: On Machine Recognition of Handprinted Chinese Characters by Feature Relaxation. In: Pattern Recognition 21.1 (1988) 1ff

[Yaj 81] Yajima, S.; Goodsell, J.L.; Ichida, T.; Hiraishi, H.: Data compression of Kanji character patterns digitized on the hexagonal mesh. In: IEEE Trans. PAMI–3 (March 1981) 221–230

[Yam 82] Yamamoto, K.; Rosenfeld, A.: Recognition of Handprinted KANJI Characters by a Relaxation Method. In: Proceedings of Sixth International Conference on Pattern Recognition, Washington D.C., 1982 (Vol.1) 395ff

[Yam 84] Yamamoto, K.; Yamada, H.; Saito, T.; Oka, R.: Recognition of Handprinted Characters and Japanese Cursive Syllabary. In: Proceedings of Eighth International Conference on Pattern Recognition, Washington D.C., 1984 (Vol.1) 385ff

[Yam 86] Yamamoto, K.; Yamada, H.; Saito, T.; Oka, R.: Recognition of Handprinted Characters in the First Level of JIS Chinese Characters. In: Proceedings of Eighth International Conference on Pattern Recognition, Washington D.C. (1986) 570ff

[Young 86] Young, Tzay Y.; Fu, King--sun (ed.): Handbook of Pattern Recognition and Image Processing. Orlando etc.: Academic Press 1986

[Z+Y 87] Zhang, Xinzhong; Yang, Deshun: A Feature Point Methode of Chinese Character Recognition Proceedings of 1987 International Conference on Chinese Information Processing (vol. II) 300ff

[Zha 79] Zhang Dihua (ed.): Xiandai Hanyu. (Modern Chinese). Hefei: Anhui Renmin Chubanshe 1979

[Zha 82] Zhang Shouxuan: A Chinese Character Recognition System Based on Pictorial Database Technique. In: Proceedings of Sixth International Conference on Pattern Recognition, Washington D.C., 1982 (Vol.2) 780ff

[Zhang 87] Zhang Zheng: Shibie xianzhixing shouxie hanzi de jiegou fenxi fangfa. (A structural analysis method for constrained handprinted Chinese characters recognition. In Chinese, English abstract). In: Beijing Gongye Xueyuan Xuebao [J. of the Beijing Inst. of Technology] (1987.1) 1–7

[Zhang 88] Zhang Xinzhong; Yan Changde; Liu Xiuying: The Feature Point Method for Chinese Character Recognition and one Application. (In Chinese). In: J. of Chinese Information Processing 1.3 (1988) 13–19

[Zhao 87] Zhao, Ming: Two-dimensional Extend Attribute Grammars for the Recognition of Hand-printed Chinese Characters. In: Proceedings of 1987 International Conference on Chinese Information Processing (vol. II) 322ff

[Zhu 87] Zhu, Xianing; Wu, Youshou; Ding, Xiaoqing: Extracting Radicals of Printed Chinese Characters. In: Proceedings of 1987 International Conference on Chinese Information Processing (vol. II) 316ff

List of Figures

(Figures where no source is indicated have been produced by the authors).

Index